Media Design

The Practice of Communication Technologies

Media Design

The Practice of Communication Technologies

Dr. Jacqueline M. Layng

Terre Layng Rosner, M.F.A.

Prentice Hall
Upper Saddle River, New Jersey 07458

Library of Congress Cataloging-in-Publication Data

Layng, Jacqueline M.
Media design: the practice of communication technologies / Jacqueline M. Layng, Terre Rosner.
p. cm
Includes bibliographical references and index.
ISBN 0-13-061028-3
1. Telecommunication. 2. Multimedia systems. 3. Mass media. I. Rosner, Terre. II. Title.

TK5101.L367 2004
621.382—dc21

2002037078

Publisher: Natalie Anderson
Executive Editor: Steven Elliot
Assistant Editor: Allison Williams
Manager, Production: Gail Steier deAcevedo
Manager, Production: Lynne Breitfeller
Manufacturing Buyer: Lynne Breitfeller
Director of Marketing: Sarah McLean
Marketing Manager: Emily Knight
Marketing Assistant: Danielle Torio
Printer/Binder: Hamilton
Cover Design: Bruce Kenselaar
Cover Illustration: Terre Layng Rosner
Composition/Full Service Project Management: Carlisle Communications, Ltd.

Pearson Education LTD.
Pearson Education Australia PTY, Limited
Pearson Education Singapore, Pte. Ltd
Pearson Education North Asia Ltd
Pearson Education Canada, Ltd
Pearson Educación de Mexico, S.A de C.V.
Pearson Education – Japan
Pearson Education Malaysia, Pte. Ltd

10 9 8 7 6 5 4 3 2 1
ISBN: 0-13-061028-3

CONTENTS

Preface ix

CHAPTER 1

Media Harmony: Pieces of the Puzzle 1

Media 3
Introduction to Communication Technologies 6
Introduction to Visual Communication 10
Merging of the Two Fields 14

CHAPTER 2

"The Medium Is the Message?" 16

Content versus Technology 19
Information in the Twenty-First Century 22
The Whole Is Greater Than the Sum of the Parts 23
The Rise of Critical Viewing 23
Critical Viewing Skills 24
The Producer and the Mediated Message 25
Critical Viewing Analysis Framework 27
Critical Viewing Model 28
Less Is More 29

CHAPTER 3

Perception: How to "See" Using Technology 32

Visual Organization 34
Visual Perception 35
Balance 36

Figure/Ground 39
Grouping 40
Repetition 41
Proximity 41
Alignment 42
Color Theory 42
Physical Color Properties 43
Psychological Color Principles 44

CHAPTER 4

Visual Technology: The Gutenberg Shift of Visual Technology **47**

Graphic Communication 48
Media Arts 49
Print Technologies 50
Desktop Publishing 51
Basic Software Needs 52
General Media Arts Software 52
Graphic File Formats 61
The Art of Visual Communication 63

CHAPTER 5

Audio Technology: Do You Hear What I Hear? **65**

Sound Communication 65
Audio Production 66
File Formats 71
Basic Software Needs 72
The Art of Audio Communication 75
Audio Designers 75
The Future 77
Streaming Audio and Event-Based Audio 78
Emerging Audio Technology Examples 78

CHAPTER 6

Digital Video: The Eyes Have It 80

Communicating in Motion 80
Composition and Continuity 81
Preproduction 82
Video Production 84
Postproduction 86
File Formats 89
The Use of Video in Presentations 90
The Use of Video in Television/Films 91
The Use of Video on the Web 91
Streaming Video 93
Copyright 94
Video Producers 95
The Future 95

CHAPTER 7

Digital Presentations: Can You Do What I Do? 96

The Definition of Multimedia 96
Presentation Style 97
Gestures 98
Interaction 99
Multimedia Elements 99
Screen Design 100
Text 101
Color 102
Illustrations 103
Charts and Diagrams 103
Photographs and Scanned Images 104
Balance 105
Sound Design 106
Motion Design 107
Presentation Software 109
Presentation Hardware 109

Display of the Presentation 110
The Art of Digital Presentations 110

CHAPTER 8

Graphic Design for the WWW: Do You See What I See? 112

Print Publishing versus Web Publishing 112
Preliminary Questions 115
Site Maps and Storyboards before the Code 117
Design and Layout Consistencies 120
HTML Coding versus Web Editors 122
Graphics and Images 122
Dynamic Content 124
Basic Web Design Rules 126

CHAPTER 9

Default Design: Garbage in, Garbage Out 128

Junk Communication 128
Communication Technology Mistakes 130
Content 130
Color 131
World Wide Web 134
Image Capture and Audio/Video 137
Text and More Problems 140

CHAPTER 10

New Technology: A Need—a Design—a Selection 143

A Need 144
A Design 145
A Technology Selection 147
Putting It All Together 148
Media Design Communication 149

Communication Edge Technologies 150
Conclusion 152

Glossary of Technical Terms 156

Index 165

PREFACE

In this sometimes overwhelming age of information, the use of new technologies in communication has virtually exploded. *Media Design: The Practice of Communication Technologies* discusses

- the current trends in the implementation of new technologies in communication and
- models of how to manipulate these many technological changes effectively in designing successful mediated messages.

In the past fifteen years, we have seen technology take college instructors from using chalkboards to manning digital power projectors. Moreover, the Internet enables communicators to tap its watershed of data instantaneously and subsequently pour out the information into meaningful output. This book applies known and accepted design concepts to varied new technologies, providing a solid practical and theoretical platform for contemporary communication teachers and students.

The chapters

- provide both the theory and practice of technology as a communication tool.

Another salutary feature of this book is that it explains and models tried and true design principles, often surrendered to the lure of immediacy and economy. The book defines the process of creating integrated solutions by explaining new avenues available for the media problems at hand. Finally, the book

- provides practical instances and supplies examples instructors can use in class to demonstrate effective media production technologies.

For example, audio is often, at best, a neglected element when designing media and, at worst, a misused nuisance. The key to using audio technology effectively is to understand the process of producing audio from beginning to end—a difficult task, since audio technology is being updated on a daily basis. From creating streaming audio for a Web site to developing audio for a television commercial, this text explores the production path to innovative solutions. One aspect of new media this book details is crafting digital audio. Through practical instruction, it guides readers to a better understanding of sound communication today.

The world of visual communication has undergone major changes in the past fifteen years—primarily, from analog to digital. Graphics, photos, and video that once took relatively extended production time to develop can now be recorded, rearranged, and displayed at accelerated speeds. New visual communication technology puts experts' tools in the hands of novices and enables laypeople to play graphic designers, videographers, editors, directors, and producers. Not unexpectedly, the quality of media output has suffered, and this text

- addresses the politically thorny issues related to unprofessional production.

We contend that communication technologies are simply tools that should, in the long run, be used to help producers communicate messages. These new tools have very often become instruments that create barriers to human communication, rather than enhance it. Some say that technology tears down the walls that block communication; however, what is left standing when the wrecking ball is still?

This book

- explains how to build bridges between the old and the new.

In the end, it is a practical path to creating ways to communicate better with each other and, therefore, to creating a more civilized society.

ACKNOWLEDGMENTS

We owe a great deal of gratitude to the many people who have contributed to the making of this book. First, a special thanks goes to the reviewers: Ross F. Collins, North Dakota State University; and Susan Escobar, Northwest Vista College; Neal Holland, John Brown University; and Scott Miller, International Fine Arts College. Their comments were very helpful and aided us in making a better book. We would also like to thank Judith E. Casillo and Anita Rhodes, from Prentice Hall, who gave us guidance when we needed it and kept the process moving along. In addition, the audio and video chapters would not have been as complete without the aid of two supportive colleagues, Don Reiber and Tom Oswald. We would like to thank our family for all their love and support. Thanks to Judge John C. Layng, who always told his daughters that they could be anything, do anything, and achieve everything in an era when boys were the only ones who played hardball. Finally, a special thank you goes to Helen K. Layng, who read through many versions of this book and never gave up the hope that it would be published.

Media Design

The Practice of Communication Technologies

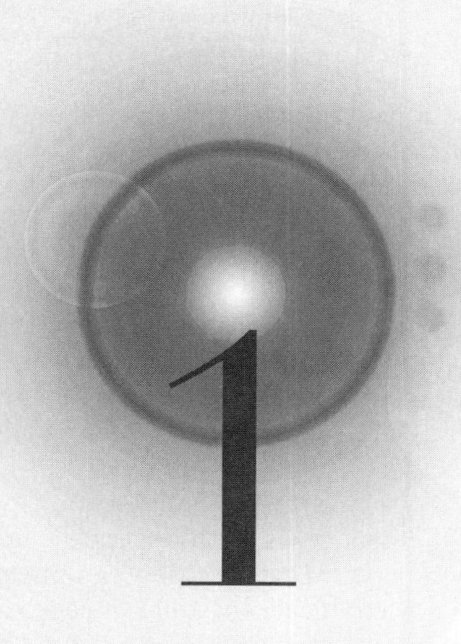

1 Media Harmony

Pieces of the puzzle

It is difficult to imagine a world without media harmony in today's technology-oriented society. There was a time when the film industry and television industry were enemies, and now they do cross-promotion of programs. So what exactly do we mean by *harmony*? For the purposes of this textbook, **media harmony** will refer to using different types of media to design effective communication. The different kinds of media include text, audio, video, graphics, animation, and Web technology. The integration of these various media to communicate has become a daily occurrence. The problem is that many people are integrating various media with little or no grounding in proven theories and practices of media design. This has created a great number of poorly designed mediated messages. It is the intention of this book to clarify what successful media design is and how to apply it to the various media utilized in today's information age. Thus, *harmony* refers to the proper use of combined media to enhance the producer's ability to communicate.

What is the "proper" use of media? Who are producers? What do we mean by "combined" media? What is communication? All are very important and complicated questions, which we intend to give detailed answers to throughout the book. We could and the field of communication has debated some of these issues for years. However, no debate is needed if proven methods are applied consistently to the use of media for communication purposes. The "proper" use of media occurs when producers employ these proven practices. In the various media industries, these practices have been used for years but are quickly being forgotten in the rage to use technology just because it is there. The Internet has come to dominate our society and has brought vast changes to the ways in which people communicate with each other. Some amazing new forms of communication have developed, but the Internet has also increased the amount of "junk" communication people encounter every day. Junk communication consists of messages that use every bell and whistle that technology has to offer and that pay little attention to the content. For this reason, it has become vital that message design play a larger role in mediated communication.

If content is important in successful communication, then the people who produce the content hold just as important a role. The producers of mediated communication come from many different fields but have one common denominator. They control how the message is designed. There are producers who control the overall content of the media, and there are producers who control parts of the media. For example, a Web site producer controls the look of the entire site and usually has graphic artists produce the visuals (such as clip art, animation, and photographs), videographers/editors shoot as well as edit the video, and Web developers produce the HTML (Hypertext Markup Language) for the various pages. If the project is small, the producer may do all the work. The traditional definition of a media producer is a person who coordinates multiple activities simultaneously and is in control of the entire media project. Producers raise the funds for the project, hire the crew, get releases, locate performers if needed, and rent or reserve facilities as well as equipment. There are also producers who are in charge of more specific areas, such as creating a video clip for the news or clip art for a Web page.

The definition of producer has not changed that much in today's technology-driven society. It has only increased the need for different types of producers, including Web developers, graphic artists, videographers, editors, writers, photographers, musicians, audio specialists, and the occasional novice. Technology and the ease of using it to create media have helped the one-time amateur become an overnight expert, or, at least, technology lets the average person have access to tools once limited to the experts. Computers and software have allowed average people and experts to become producers in many different media. A graphic artist can shoot and edit video for streaming video on a Web site. Web developers can integrate audio, video, and animation into a Web page with a few clicks of the mouse. Thus, media producers have become any people involved with creating media for consumption by the public. This includes producing for the medium of television, integrating audio and video onto the Internet, or combining the use of multiple media to create a computer-based training compact disc.

What exactly are "combined" media, and aren't all media already combined? Combined media consist of one or more media integrated into another medium—for example, using audio on a Web page or using radio with the Internet. The combined use of media has occurred for many years—for instance, the use of photographs, music, and animation in television. However, today's technology has brought about media convergence. Media that were separate entities are becoming one. It was only a few years ago that the Internet was a text-oriented medium. Today, the integration of graphics, audio, video, animation, and photographs makes it a medium of combined media. You can now get the Internet on your phone and answer your phone on the Internet. All of this is an example of media convergence. Eventually, we will be able to access all available media from a single source, such as a cable box, a phone, or an Internet provider. The obvious question is "How is this changing communication?"

In order to understand how these new technologies are changing communication, it is important to understand what communication is. The study of communication can be traced back thousands of years to Ancient Greece and the Socratic method. These were the earliest question-and-answer sessions that produced communication exchanges as well as knowledge. Since then, hundreds of theories and models have analyzed intrapersonal communication (communicating with yourself), interpersonal communication (communicating with another person), group communication (communicating with more than two people), and mass communication (communicating with many people at once). All such theories and models discuss common factors, such as the sender, encoding, the receiver, decoding, the message, transmission, and feedback, with varying themes. Most agree that communication does not take place without understanding the message. The receiver must understand the message, and the sender should understand the feedback, for successful communication to take place.

The most basic communication model consists of the sender having an idea and encoding it (such as into written language or speech), then transmitting it (such as by pen and paper or voice) to the receiver, who decodes (reads or listens to) the message and responds (gives feedback) to the sender. If this process does not take place, it is thought that communication does not occur. However, in the information age, the sender does not always initiate the process; sometimes the receiver pulls the information from the media-rich environment. In other words, once a sender produces information on the Internet, it is there for receivers to find and decode when they want to, and this is called pulling in the information. The receiver can also choose to respond or not respond and can choose how to pull in the information. It is not a linear process, as past communication has been, but, rather, can be accessed spatially. The basic principle is still that a message is sent and received as well as understood, in order to be identified as successful communication. It is the process that has changed, empowering the receiver with choices never before imagined. The sender can create a message that can be decoded by millions of receivers and can design the message so that each receiver can pull in the information in multiple ways. Also, the message can be encoded in many different ways, such as in several languages. The message can be media-rich (can use multiple media) and pushed at the receiver. The sender can collect information on the receiver when he/she accesses the message and can send other messages the receiver might or might not be interested in decoding. New technologies have increased the amount of information we encounter, the speed at which we get it, and our access to people around the world, and they have changed the way people communicate.

Communication is people exchanging information through various media to better understand themselves and the world around them. Some may say that computers exchange information and that is communication, but it is human communication that this book is concerned with and will concentrate on in discussing media design. The use and design of media by producers to communicate messages and secure certain outcomes are the focus of this book. Human communication is about bringing the receivers/users of information to a desired response. Communication can be used to inform, to persuade, to organize, to act, to educate, or even to survive. Human beings are social animals and need to communicate to survive as a species and as individuals. We use communication to find mates, friends, jobs, cars, houses, communities, and so on. Most media producers communicate to inform people of a subject or to persuade them to buy a product or to change attitudes/beliefs about a topic. Producers of media communicate for a reason, and that reason usually has to do with generating money. Human beings sometimes communicate just for the sake of communicating. The successful media producer can do both but usually participates in the former. Nonetheless, communication is the key factor in media design. The end result is media harmony with benefits. See Fig. 1-1.

MEDIA

Media is a word that is used a great deal but often misunderstood. Most people think of television and radio, which are considered mass media. These are technologies that deliver information to the masses, but does a technology have to transmit

FIGURE 1-1

A book, a radio, and a TV

information to many to be identified as a medium? To understand what the word *media* refers to, it helps to briefly trace the history of human communication. Throughout history, humans have communicated with each other by various modes, language and gestures in the beginning and later drawings on caves. These drawings can be classified as a medium. Human beings used a vehicle to communicate to current and the next generations. The paint and sticks used to create the drawings are media. Media have evolved from paintings to pen and paper, to hand-copied books, to mass communication through the invention of the printing press. Print in the West became the first mass medium with the help of Johannes Gutenberg's invention of movable type, making the printing press a viable technology. Books, newspapers, and magazines developed from this technology and are considered media. Media continued to evolve with the invention of the photograph and film. These media used images and eventually sound to communicate with the masses. Once society started to use electricity, radio and television became possibilities, as well as incredibly powerful media in the world. The latest addition to the mass media and currently thought to be one of the most powerful is the Internet.

The technology used as vehicles for transmitting messages to others has developed into the various media industries. Many inventions along the way have aided in the creation of the various media, such as the telegraph, telephone, and analog and digital signals. These inventions have paved the way for the print, electronic, and digital media industries of today. Each medium has its own intrinsic values that make it a unique medium. For example, a photograph prepared for a magazine and a photograph prepared for the Internet require very different design practices to make them successful communication. Media convergence is taking place, but that does not change the need to pay attention to the unique characteristics of each medium. Obviously, print media require higher-quality images than clip art on the Internet. Audio clips on a Web page do not necessarily require high-end sound quality, while a radio broadcast does, if it is going to be listened to on a regular basis. The characteristics of each medium affect the design of the message.

There are two factors to pay attention to when designing media. The first is the identity of the audience and the second is the end product. The audience members must dictate the design, because they are the reason for communicating in the first place. Producing a message for a select few people is different than creating a message for thousands. This affects not only the design but also the selection of the media that best fit the need of the communication situation. In other words, the Internet is not the answer to all exchanges of information between human beings. A pamphlet with photographs could promote your idea better to the audience you are trying to reach. For instance, why produce a television commercial in HDTV (high-definition television) if your target audience does not own the technology? That would be a waste of time and money, yet it happens in today's society. The moral of this story is always keep your audience in mind when choosing a medium and designing a message to be used on that medium.

It is also important to get to know how each medium works and what you want for an end product. Designing animation for a Web site is not the same as designing it for film or television. Web animations tend to be limited in size, while film animation fills the screen. The process of developing the animation may be similar, but the output requires a different result. A fifteen-minute promotional video to be displayed at a conference may not work as well on the company's Web site; the former requires different shots than the latter. As a case in point, the conference video should use wide angles and more long shots, while Web site videos need to use more close-ups and narrow angles. Also, certain questions must be asked and answered when designing mediated messages. Will the message be projected, downloaded, uploaded, e-mailed, or transmitted by wireless telephones? The end product of the mediated message has an impact on the design of the message.

Media producers that do not understand the many characteristics of the media or the identity of their audience are doomed to create junk communication, poorly designed messages that are hard to understand and cannot hold the audience's interest for long. This is why it is vital that you understand the basic definition of *medium*, which is a vehicle/technology that transmits information to other human beings. If you fail to learn how to use that vehicle, then you fail to communicate. This failure will assure you don't receive the desired outcome of informing, persuading, or moving the audience to action. This usually occurs when media producers let the technology choice rule the design, which can produce mediated messages that do not communicate well. Understanding both the technology and the content, not just the technology, should drive the design. See Fig. 1-2.

FIGURE 1-2

A computer, a satellite, and a cell phone

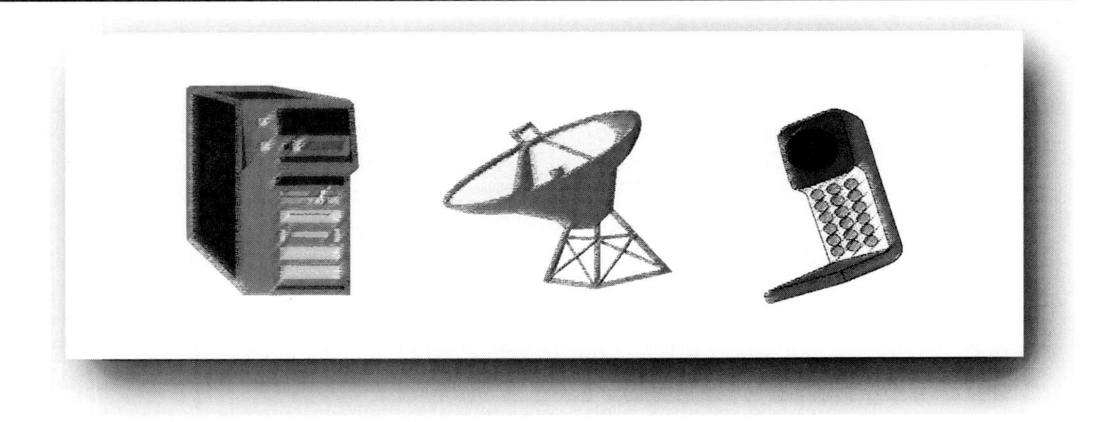

INTRODUCTION TO COMMUNICATION TECHNOLOGIES

Now that you understand more about the communication process, the media, and their producers, it is necessary to learn more about the multiple communication technologies available to become better media designers. Literally dozens of technologies make up the communication technologies currently being used by society. It is difficult to discuss these technologies without first understanding what communication technology is and why people use it. You can almost use the same definition we used earlier for defining a medium as a vehicle/technology that transmits information to other human beings. In fact, the two have practically become known as the same, depending on the field using the term. You will even find colleges and universities that use the term *telecommunication* in place of *media production* or *broadcasting*. The terms used to mean different things: *telecommunication* referred to telephone technology for transmission, while *broadcasting* referred to the electromagnetic spectrum, cable, and microwave technology for transmission. However, the delivery systems of the different fields have merged now, and the traditional media (radio, television, and film) use telephone technology and vice versa.

The different fields' delivery systems have merged for two reasons. The first reason is technology. The development of a digital system that delivers higher-quality signals improved the telephone industry and the media industry. The telephone companies could deliver more calls over smaller lines with better voice quality, while the media companies could deliver higher-quality sound and video with more choices. The second reason is the Telecommunication Act of 1996. This comprehensive legislation changed the regulations on the media, telephone, cable, and satellite industries, allowing them to have cross-ownership. It is now legal for your cable company to provide Internet access, telephone service, and television programming to its subscribers. The telephone companies have begun to buy media stations, so that they can provide several communication services to their customers. Media companies are buying up Internet service providers, radio and television stations, film production companies, and telephone systems to offer a comprehensive service to their customers, and satellite companies are doing similar things. The Telecommunication Act of 1996 made the AOL/Time Warner merger possible, which created the largest media conglomerate to date.

The implications of these growing media conglomerates on society are currently being analyzed, and the jury is still out on the possible effects. On one side of the debate are those who said that these companies will provide one-stop shopping and the lack of competition should keep the cost down. The opposing viewpoint is that a few companies will control all the information going to the public and, thus, manipulation of people will be easier. Nonetheless, this media convergence has basically made media technologies and communication technologies synonymous. This is the age of information, and people use these communication technologies for working, playing, meeting people, and communicating with family and the world. Our society has become globally linked through technology, and it is vital for people to understand the tools we use to communicate with each other, or they can and will be manipulated.

A good defense against being manipulated and becoming better communicators, not to mention improved media producers, is to get familiar with the various communication technologies used in society. Let's begin by looking at the cable and telephone industries and the current technologies in use. The cable and telephone industries have been busy rewiring their systems with fiber-optic technology. This technology has aided these industries by increasing speed and volume. A fiber optic is a glass rod that uses light to transmit a digital signal to various points. These glass strands are capable of carrying data as light and were invented by Corning Glass. The strands carry more data than copper wire. Copper wire was the old cable and phone system of carrying analog data. A single fiber-optic line can carry 60,000 telephone calls simultaneously. This improved technology has created a host of new communication services the cable and telephone companies can offer to consumers.

The cable industries have been rewiring the new system for the past few years, and this has let them jump ahead of their competitors. The Federal Communication Commission (FCC), which regulates the media industries, set a deadline of going digital by May 2006. This deadline means that all television and cable channels must be in digital form by that date. The cable industry is well on its way to meeting the deadline because of its new fiber-optic infrastructure. Cable systems are now offering digital set-top boxes, which take digital television to subscribers. The new services offered are video-on-demand, e-mail, and Web browsing through the TV. Video-on-demand allows cable subscribers to download data to a set-top box to watch a particular movie whenever they feel like it. They can stop the movie and go back to watch it hours later, and it is at the exact spot in the movie they left it at. Sony has developed a set-top box that allows subscribers to check e-mail and surf the Web while watching television. Currently, cable companies offer access to the Internet with cable connections that bring data to your computer at a cost-effective and faster rate than the old dedicated telephone lines, such as the T1 phone line.

However, the telephone industries have not been left too far behind in the conversion to the new digital system. The Telecommunication Act of 1996 gave these companies the ability to buy cable and television stations. This is allowing the telephone companies a chance to branch out and offer services similar to those of the cable industries. In addition, it has brought down the cost of long distance calls and has opened up a whole new market of Internet access. Telephone hook-ups and modems have been the number one way Internet users have accessed the Web. In the past, telephone access has had its difficulties, with busy signals, cutoffs, and slow speed. The new digital system has increased the quality of service, but cable access is gaining popularity. The telephone companies have been fighting this with added services, such as caller-ID, voice mail, and so on. It appears that telephone services and cable services are headed in the same direction to provide all the information and entertainment entering the home.

The television industry is in the same predicament as cable and must become digital by 2006. This industry is having more problems making the change than cable. Television until recently has been using the analog system to record and transmit a signal to viewers. The analog system sends a signal of electrical impulses through the electromagnetic spectrum to television sets. The current analog signal has been converted to a digital signal to be transmitted by satellites and cable industries. The problem the television industry is having is converting the analog equipment to digital equipment in more than 1,400 stations. The cost has been high, and fewer than half of the stations currently offer digital transmission in addition to their analog signals. The industry is having other difficulties, such as HDTV, which is a hybrid analog and digital system that delivers a higher-quality picture. Do the stations offer HDTV, DTV (digital television), or both? The answer to that question can cost each station millions of dollars in equipment. Then, there is a digital system that has been approved that was developed by the Grand Alliance, a group of media companies interested in creating a digital television system. The TV industry is still debating some of the standards to be set for the digital system being adopted.

There are some other problems with converting to the digital system. A digital video format standard has not materialized. Not one digital equipment system has become a standard for recording and editing programs. There are many digital formats to choose from, including DV, DVCAM, Digital BetaCam, DVCPRO, and other, similar formats. There has not been a standard selected for digital receivers, and this is causing further delay in the 2006 plan. Broadcasters do not believe they will make the 2006 deadline and are looking for an extension. Nonetheless, digital television and video are a reality and currently being used in various ways. Digital video has made shooting and editing video easier and more cost-effective. The picture quality has improved, and it can be converted easily to the analog system or integrated into a Web site. Broadcasters have been offering enhanced TV, which combines regular television programs with the Internet. Accompanying each TV program is a Web site that provides more information to the viewer and may or may not be synchronized with the TV show. The development of CNBC, MSNBC, AOLTV, and others displays a merging of content between television and the Internet.

The Internet—which evolved from ARPA (Advanced Research Project Agency), U.S. Defense Department computer network that preceded the Internet, and the development of the semiconductor chip, which made digital technology possible as well as compact and other technological advances—has become a highly popular communication technology. This new technology has made immediate interactive communication across great distances at a low cost available to the world. The Internet, which was developed by the government and researchers to exchange information among a select group, exploded onto society's radar when it was opened to the public in 1992. This technology has brought more information and communication to the public than any other technology. The Internet has empowered individuals and has created whole new industries. It has also helped merge the various media by offering telephone services, videophone, television broadcasts, radio broadcasts, movies, and recording artists through one medium. The quality is not always the best, but there is access to media-rich information that was not possible ten years ago. Quality is also improving with compression software and high-speed access increasing with new developments on an almost daily basis. For example, you could not watch television or radio broadcast on the Internet a few years ago, and now there are Internet radio and TV stations.

Teleconferencing and computer conferencing have also become much easier with digital technology. It used to cost a great deal of money to send a two-way television signal between two sites. This was known as teleconferencing; however, with the Internet, the technology has been made easier and costs much less to produce. Today, organizations can put cameras on their computers and, with the right software, see and talk to one another in real time. The pictures are smaller than a regular full TV screen and the audio is slightly delayed; however, as compression and access speed increase, so will the quality of computer conferencing. Computer conferencing is also being used more because of advances in computers, such as processing speed, RAM (random access memory), and modems. These improvements have made many aspects of the Internet a viable alternative to the technologies that originally provided similar services. For instance, digital radio and music have also become very successful parts of the Internet. Traditional radio stations all have a Web presence, and many Webcast their programming in addition to airing regular broadcasts. Radio stations are finding new areas of revenue and can sell the music directly to the consumer. There have been some setbacks to digital radio, however, Why listen or buy music from the stations or the recording industry when you can get it for free? A company called Napster developed a software program that made sharing audio files easy and threatened to make the recording industry, compact discs, and all other playback systems that were not digital obsolete, even with the latest legal developments. A successful lawsuit by several recording artists and studios against Napster caused its end, but the use of the technology has continued. For instance, portable devices (such as MP3) that download music directly from the Internet are now available, and digital radio in your car is just around the corner. Several companies (Ford and Nasa Neural Network) are testing it and using wireless technology.

Wireless technology is basically technology that is not directly wired to the sending and receiving technology. This technology has also experienced a surge in use with the public. The key to this technology is sending signals through the air by digital towers and satellite transmission. Direct broadcast satellites have been used for years to send voice and data around the world but only recently have come to be used for specific purposes. DirecTV is an example of using satellites to transmit television channels. The digital phone systems use satellites to send signals, and the automobile makers (GM's Intelligent Vehicle initiative) use them to send data and radio signals to and from cars. It is this technology that is making it possible to access the Internet through your mobile phone and may eventually be how we access all information. It is basically text-oriented information being sent over mobile phones now, but it will expand as the technology improves the power of the signals and the infrastructure. The future of wireless technology could hold an entirely different lifestyle for us. This technology could aid in media convergence, and every appliance, entertainment, and communication device could be programmed and controlled from a distance. In other words, telephone calls made for you while you're away, meals cooked, atmosphere adjusted, and only your favorite television programs waiting for you because they have been downloaded to your location by technology devices. Your car will play only the digital radio stations you desire, and you don't have to listen to "talk radio" if you don't want to. This is a form of artificial intelligence (AI).

Artificial intelligence is the ability of a computer to think as a human being and to make choices. It is believed that a computer will be able to be programmed to think and act as specific human beings. For example, you may be able to program your house with your personality, and the house would make the same choices for dinner and a movie you might make for yourself. The idea is that computers can be programmed to learn and react as the human brain would react to a similar situation. This technology is currently being developed and may hold the true key to media convergence. This is all information coming from one technology that is controlled by that technology.

Finally, there is virtual reality (VR). This technology allows a computer to simulate reality with three-dimensional images. It is being used to build houses and to model cities that do not exist. VR has been a popular form of entertainment. People put on helmets and gloves and become a part of the computerized environment. VR continues to be developed and used for purposes ranging from training surgeons to creating new video games. Entire suits with sensors have been constructed to allow the persons wearing them to feel sensations taking place in the virtual world. VR has not progressed as fast as first expected and still has a long way to go, for example, before we are on *Star Trek's* holodeck. Still, VR and AI are progressing and being used with other technologies, such as film and television, and it is only a matter of time before these technologies are more commonplace.

These are just a few of the major communication technologies being used in today's society. Every day, these technologies are being updated and refined. Technology is always changing and progressing as societal needs change and progress. There are so many levels to each of the various technologies, and it seems almost impossible to fathom all the possibilities that exist with the varied uses of technology. It is for this reason that this book focuses on the media design of various technologies. On the most basic level, these technologies are delivering information that requires a certain structure if people are going to understand it. Any good media designer keeps this in mind when crafting a message to be used on one of the technologies mentioned earlier in this chapter. We will attempt to minimize the confusion and focus our attention on technology that is an audio/visual medium, such as the Internet or digital video.

Paintbrushes and a palette

Introduction to Visual Communication

Visual communication is the practice of using imagery to convey information to a viewer. The generally accepted modes of visual communication are those elements that a human can "see"—for instance, graphics, eye movement, and body language. We communicate in many different modes. Verbal communication is only one method. In fact, most communication is sent by other means. An example of this is the simple children's game Simon Says. As the sender commands an action from a receiver, the receiver attempts to react to the message as directed by the sender. The interesting aspect of this game is that, in most instances, the receiver is fooled into reacting incorrectly because of what he/she sees, rather than what he/she hears. The power of visual communication is unmistakable. Throughout human history, communicating by using graphic imagery has been evident. Even our earliest ancestors were aware of these powerful communication tools.

The milestones in the history of visual communication stretch back at least 40,000 years. The earliest communication with images can be traced to 35,000 B.C. in the early drawings discovered in the cave paintings of Vallon-Pont-d'Arc in southeastern France. By 3800 B.C., in what was considered ancient Sumeria, the first graphic icons showing early written communication had appeared in the Blau Monument. Approximately 1450 B.C., the invention of papyrus, or paper, emerged, demonstrated in the ancient Egyptian *Book of the Dead of Tuthmosis* III. These early inventions in communication were the driving force for further discovery in the technologies to distribute visual information to an audience. At this point in history, the distribution of graphic information was painstakingly, individually developed by hand. Graphic communication was enjoyed only by the privileged few.

At the turn of the first millennium A.D., the Chinese invented methods of actual block printing and the use of movable type for printing. During the Dark Ages, around A.D. 800, European monks were creating amazingly beautiful illuminated manuscripts with gold embossed text and illustrations. In the next 500 years, Renaissance inventors developed and perfected the printing press and typography in Germany. The explosion of mass-distributed graphic communication was inevitable, and this general distribution of imagery and text changed the way the world communicated. Before the invention of printing technologies, people gleaned their information from sparsely distributed, hand-created graphic communication. With the advent of "publishing," masses of people became able to read and interpret the same printed information, thereby switching to communication through mass sources, rather than individual sources. The first technology trend in visual communication was in full swing. See Fig. 1-3.

Once the logistics of and practice with these new printing technologies became commonplace, the knowledge that printing can be used to inform, inspire, and persuade emerged. Methods of "advertising" were tried through the invention of the poster and lithographic printing methods during the industrial revolution. With the invention of photography by Louis Daguerre in the 1830s and motion photography by Thomas Edison and the help of the aperture tripping experiments of Eadweard Muybridge in 1880, the first technology trend in visual communication was ending while the second was beginning. Admittedly, the invention of the printing press revolutionized written communication, but, likewise, the invention of photography forever transformed graphic communication. The ability to mass produce nearly perfectly realistic imagery shifted the focus of visual communication from text to pictures. Photography solved the dilemma of requiring highly trained craftsmen to create unique illustrations. Potentially, photographs could communicate more effectively than hundreds of words. The ability to mass produce both text and images and to distribute the results to an eager audience was possible. The form of graphic communication drastically changed. See Fig. 1-4.

Thus far, this discussion of visual communication history has focused on the technological advancements in the form of graphic development. Strictly speaking, the technology by which graphic communication was created and distributed was relative to the content that was conveyed. People's customs, state mandates, and visual elements varied the content of these communications. Therein lie the two elemental aspects of visual communication, form and content. Inherent in the discovery of designing printed communication to persuade an audience came the ability to analyze this process. The producers of visual communication, both textual and pictorial, were naturally drawn to questioning the reasons their audiences were motivated by certain visual stimuli in the form and content of text and images.

By consciously applying basic visual principles of design perception and ultimately communicating with an audience, a media producer can very often prompt a desired reaction. This persuasion power drives most graphic communication

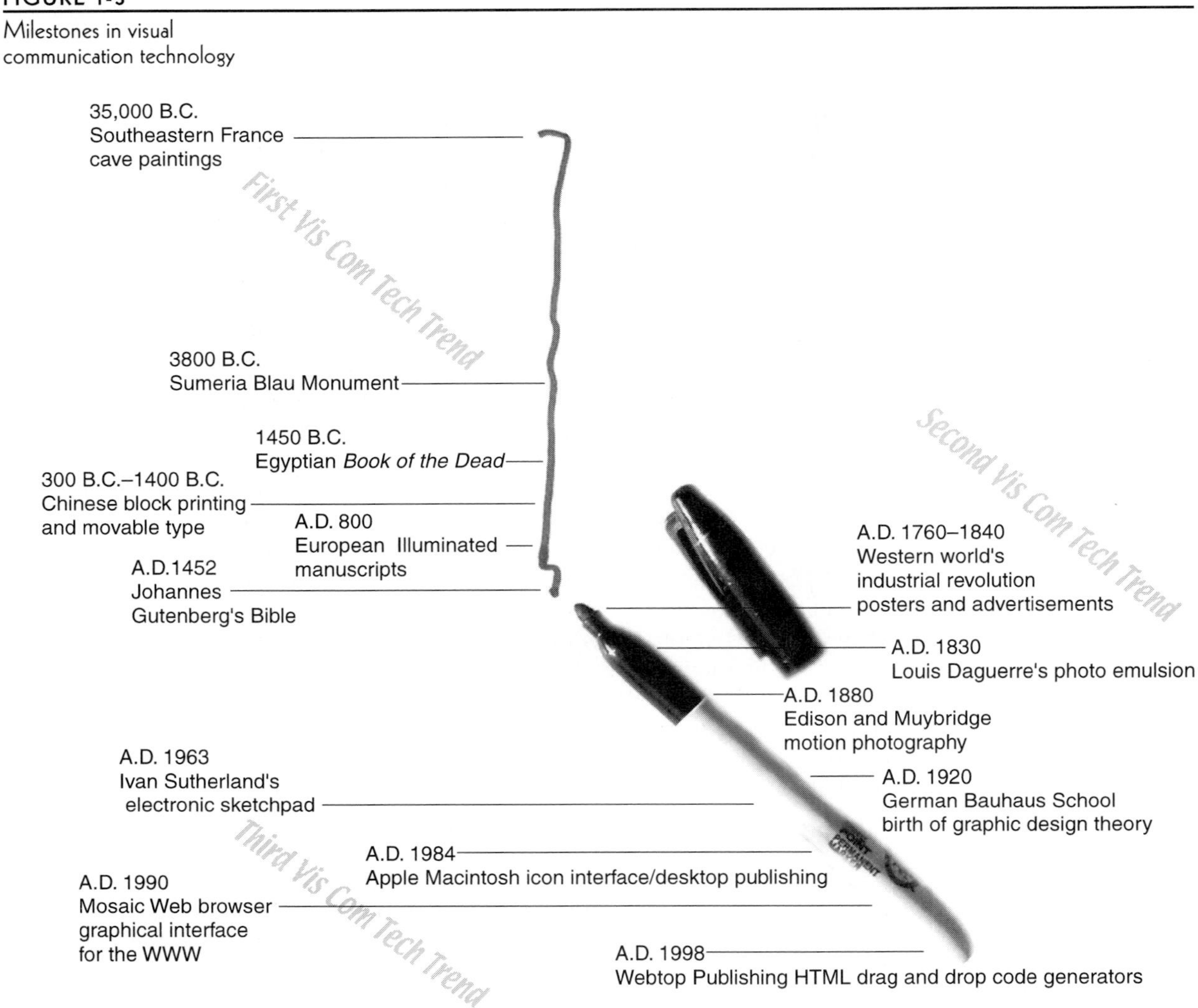

FIGURE 1-3
Milestones in visual communication technology

developed in today's commercial arts industry. The ability to move an audience to action is at the heart of effective visual communication. For thousands of years, artists in the field of fine art have controlled the creation and interpretation of images, and eventually areas of specialization have emerged. One of those areas of specialization is graphic design. Graphic designers have been creating intentional images for commercial media for sometime. Some experts argue that graphic design, the intentional persuasion of an audience through visuals, has been in use for 30,000 years. Others believe that graphic design could only have been in existence since the development of Gutenberg's printing press around 1450, the first real mass publishing medium. Still others argue that this field could not have been in existence until the producers of these visual images were conscious of the power of their communication techniques. In other words, if artists were unaware of the persuasive impact of their work, graphic design did not exist. According to the last explanation, the birth of modern

FIGURE 1-4

This "tintype" is an early example of photo emulsion exposed on a tin backing rather than on glass, as in daguerreotypes

graphic design theory, defined as the layout of text and images with the intention of communicating a specific message perceived by an audience within given limitations, began in the1920s during a period in art history labeled "Bauhaus." The Bauhaus was a physical place where artists, writers, philosophers, psychologists, and academics set out to define how and why humans organize the elements and spaces they perceive. These intellectuals came up with a series of explanations, definitions, and theories of basic design perception. In order to produce effective graphic design, the producer must design by intention rather than by default. The creator must understand why the practice of developing and arranging visuals to convey a specific message or visual communication is not successful. Any element within a field of space has the potential to convey a certain message. The Bauhaus theories defined how an artist can virtually control the way a viewer will generally interpret these elements.

Traditionally, graphic design has been restricted to a two-dimensional medium. Graphic designers have developed images for books, magazines, television, film, and newspapers. They have been producing illustrations, photographs, marks/logos, products, exhibitions, industrial, architectural and TV graphics for the past eighty years. As the field matured, an entirely new area of graphic design was born in 1963, when Ivan Sutherland invented sketchpad, the first interactive computer graphic. With the invention of the personal computer and Macintosh's icon-driven interface in 1984, the third technology trend in visual communication began. Mirroring the process of making visual communication technology commonplace, as in the first trend, computer programmers began developing software packages that allowed graphic communicators to create published communications without having to master the underlying code. Like an early-sixteenth-century printer who used a printing press without having to build the mechanics of the press, the graphic artist was freed from having to write the programming code to produce a computer-based layout. Because the technology had a relatively low learning curve, desktop publishing by even the most novice user of personal computers was possible. This computer-driven mode of developing printed materials created a new niche for graphic communicators. Computer graphic designers make clip art, icons, logos, CAD/CAM, animations, Web pages, Web graphics, and interactive interfaces for the Internet, as well as computer-authoring programs.

The invention of the Internet as an experiment by the U.S. Department of Defense as a way to communicate with scientists and academics in the 1960s continued the third technology trend in visual communication. The Internet, which at its base is simply a vast, worldwide group of networked computers, was initially very difficult for a layperson to use. Driven by the need to lower the learning curve in the use of the Internet, a group of European scientists at CERN, a European organization for Nuclear Research, invented a way to access information on the Internet, called the World Wide Web (WWW). Following a familiar pattern, Internet technology became substantially easier to use in the early 1990s, with the invention of Mosaic by Marc Andreseen, a browser using the first graphical interface for the WWW. Taking a cue from visual communication history, producers on the Web immediately realized the potential of this new communication medium, causing an explosion in the need for computer graphics. As the Web becomes more visual and interactive, the need for computer graphics will continue to increase. The development of software programs that aid in making print collateral and Web publishing has helped graphic designers in producing visual communication. These software programs range from the easy-to-use templates of an HTML composer in a Web browser to the more complex, stand-alone Web editor such as MS Front Page®. There are many other software programs used to develop visual communication for the Internet and digital media. Media designers also use nonlinear video editing and special effects programs, image-editing and manipulation programs, bitmap and vector illustration programs, bitmap and vector animation programs, and streaming content creation programs. New programs are being developed and at various user levels. Professional media designers commonly use dedicated, multifaceted, complex programs with high learning curves, while amateur producers tend to use all-inclusive, generalized, user-friendly programs with low learning curves. For instance, professional Web designers use Web editors that allow for raw code and script editing in tandem with drag and drop publishing elements. On the other hand, word-processing-like interfaces integrated into common Web browsers are the preferred Web composing program selections of a novice producer. The list of software programs available to produce digital media is almost endless and continues to change as visual communication technology is updated.

MERGING OF THE TWO FIELDS

An interesting phenomenon is occurring in the field of communication; visual communication, which traditionally has been studied in the area of broadcasting and journalism, has matured into its own area of study. Once considered to be the domain of the arts, visual communication has been born anew in the field of communication. In fact, more and more colleges are offering visual communication and communication technology as majors. Art students are beginning to study communication, and communication students have started to study art more closely. The two fields are undergoing a metamorphosis, and the results are new jobs for students who embrace these changes. A few years ago, broadcasting students went to work at broadcasting stations and produced news and entertainment programs. Graphic art students designed commercial art for magazine, book, and newspaper publishers; television; and private organizations. Now both art and communication students are working for Internet, cable, telephone, television, wireless, and other media conglomerate companies, developing audio/visual communication. They have become the workforces of the present and future.

Many career opportunities exist for students who study the design of new media technology. The traditional careers in computer graphic design, broadcasting, and journalism still exist, with some new twists. There are more opportunities for work since the explosion of media outlets, such as Internet companies, a Web presence for all traditional media organizations, Web-only media, and brick-and-mortar corporations that use communication technologies for normal business activities. A good designer of media technology may create whole Web sites or the audio and video on Web sites or may produce content for .com companies. The possibilities are endless for students who study and learn the proper techniques of media design.

These are the various reasons that understanding good media design theories and practices have become important to study. Media harmony comes from thoroughly knowing the media you are working with and having the ability to design a visual message to convey meaning to an audience through the use of multiple communication technologies. This book focuses on creating media harmony by discussing the medium as the message; perception; visual technology; audio technology; digital video; digital presentations; graphic design for the WWW; default design; and the outcome of properly designed new technology.

BIBLIOGRAPHY AND SUGGESTED READING

Arnason, H.H. (1981). *History of Modern Art*. New York and Englewood Cliffs, NJ: Prentice Hall and Harry N. Abrams.

Arntson, Amy E. (1998). *Graphic Design Basics*. Orlando, FL: Harcourt Brace.

Berryman, Gregg. (1990). *Notes on Graphic Design and Visual Communication*. Menlo Park, CA: Crisp.

Donahue, Bud. (1986). *The Language of Layout*. New York: Prentice Hall Press.

Meggs, Philip B. (1983). *A History of Graphic Design*. New York: Van Nostrand Reinhold.

Mirabito, M. (1997). *The New Communication Technologies*. Boston: Focal Press.

Powell, Ann. (1973). *The Origins of Western Art*. London. Thames and Hudson and Harcourt Brace Jovanovich.

Rosenblum, Naomi. (1984). *A World History of Photography*. New York: Abbeville Press.

Scharf, Aaron. (1974). *Art and Photography*. Kingsport, TN: Kingsport Press.

Vaughan, T. (1994). *MultiMedia: Making It Work*. Boston: Osbourne McGraw-Hill.

Vivian, J. (1999). *The Media of Mass Communication*. Berkeley, CA: Allyn & Bacon.

2 "The Medium Is the Message?"

Marshall McLuhan's famous quote sparked a debate that continues today. The increased use of communication technology in society has only intensified the debate. What is the message—the content, the technology, communicators, or a combination of factors? To understand where the field of communication technology is headed in the future, it is important to revisit past research. In the age of information, the use of new technologies has virtually exploded, and media producers are scrambling to keep up by using these technologies at an increasing rate. Thus, it is important to understand how to integrate this new technology into communication. Research and theorists already led the way in understanding the use of this technology in communication. The history of communication technology pedagogy is vast and complex and encompasses theories that have been embraced by many disciplines. Some trends evolving in the field today can be traced back to several important theorists who set the stage for the new age of communication technology—an age built on the use of ever emerging technologies to aid communication. These theorists reflect a paradigm shift from analyzing media hardware to discovering the role of communication and the place of the audience in that process. Revisiting these theories will help provide a foundation for the use of present and future technologies in the design of mediated messages.

In this chapter, several important theorists who have made and are still making a major impact will be discussed. The first half of the chapter will be an overview of four theorists, Jean Piaget, Robert Gagné, Leslie Briggs, and Gavriel Salomon, all of whom have had a great impact on the course of communication technology. The second half of the discussion will be a more in-depth look at Gavriel Salomon's major effect on and the future of this ever evolving field. "It should be seen as a system of practical knowledge not necessarily reflected in things or hardware" (Saettler, 1990, p. 3). Basically, technology is more than the tools that producers use for communication; it is a process of molding the environment of an audience to increase knowledge or to persuade. These four theorists have made quite an impact on the field of communication, as well as on other fields, such as educational psychology. Their theories have aided in shaping the environments of many varied fields. In fact, Piaget, Gagné and Briggs concepts have been so influential that it is hard to imagine the field of communication technology without them. Their theories analyze viewer development and systematic communication. It is obvious that hardware cannot be an effective tool if the producer is unaware of the viewer's capabilities and designs the incorrect system of communication. There are many factors to take into consideration in media design. Piaget, Gagné, and Briggs analyzed two important factors in the communication process, cognition (knowledge) and systematic communication, a series of steps that aid in the transfer of information.

Piaget believed that cognition develops from contact between *a message* receiver and the environment. Two terms from his cognitive theory describe this relationship: *assimilation* occurs when the receiver copes with a situation that was originally too difficult, while *accommodation* is the receiver's adjusting to a new environment and adding this new information into his/her scheme. Piaget believed the scheme was a process of creating balance between the receiver and his/her environment. Piaget questioned these different processes (assimilation, accommodation) of change by analyzing the development of cognition. Piaget envisioned four stages of cognitive development: sensorimotor, preoperational, concrete operations, and formal operations. These four stages of development occur in the cognitive growth of an individual. Piaget was attempting to understand the cognitive process through experimentation with children. His theory provided new approaches, as opposed to old ones that looked only at receiver maturation and quantity of brain activity. This is a perspective about the audience, which includes the receiver's environment and cognitive abilities.

How is Piaget's theory influential in the field of communication technology? It provides a scientific perspective on the cognitive development of an audience. By understanding the mind of a receiver, a media producer can create better designs and use the correct tools. Piaget's theory gives producers a basis on which to build knowledge. It allows them to develop systems of designs by looking at the individual's internal scheme in connection with his/her environment. Perhaps this is why Gagné and Briggs' theory of instruction appears to be an extension of Piaget's theory of cognition.

Gagné and Briggs developed the theory of instruction through a compilation of their previous work. The theorists support their concept of instructional design by making five steps, which can be applied easily through a communication perspective. First, aid the communication of the individual; second, both intermediate- and long-range goals should be planned; third, communication should be systematically designed; fourth, communication should be conducted by means of a system approach; fifth, design should be based on how human beings communicate. This process systematically focuses the message on a particular audience with maximum impact. All five steps deal with addressing the needs of the receiver through the design of the message.

It is important to note that Gagné and Briggs analyzed old and new principles of learning theories. According to Gagné and Briggs, there are two factors involved in a communication event: (1) external factors that use contiguity, repetition, and reinforcement and (2) internal factors that use factual information, intellectual skills, and strategies. These factors are principles that a producer can also use to aid in the communication process. Gagné and Briggs thought that human communication/learning is a set of internal cognitive processes that transform the receiver into several successive phases of information processing. The result of this information processing is communication, which consists of intellectual skills (aid the receiver in carrying out symbol-based procedures), cognitive strategies (aid the receiver in cognitive processing), information (facts are organized and stored in memory), motor skills (the physical activity required for purposeful actions), and attitudes (modify an individual's choices of action). Although Gagné and Briggs' theory is based on learning, there is a definite connection to the communication process and their theory flows effortlessly into understanding and better designing the act of communication.

Using Gagné and Briggs' theories, successful communication should occur if a series of steps take place in approximately the following sequence: gain the receiver's attention, inform the receiver of the objective, stimulate the receiver's memory, present the mediated message, elicit receiver performance, provide feedback, and assess the receiver's feedback. This sequence is defined as communicating. It is evident that one can make a parallel between Piaget's theory and Gagné and Briggs' theory. Both theories take into effect external and internal conditions, although they analyze the receiver from different perspectives. Piaget is defining the receiver, while Gagné and Briggs are developing a system to inform the receiver. The latter theory elaborates further by systematically creating a communication design model.

This system of designing communication encompasses Gagné and Briggs' earlier discussions of human capabilities. It organizes communication and provides models that can be applied in multiple situations. Through the use of these stages, a media producer can systematically present information to an audience. The design model allows a producer to assess the receiver and his or her environment. It is a complicated theory, which analyzes every aspect of communication. Nonetheless, the positive aspects of Gagné and Briggs' theory of instruction have been a major influence in the field of communication technology.

A variety of theories, such as humanistic, psychological, and behaviorist theories, have affected the field of communication technology. Many have been influential in shaping the field, but few have made the unique impact of Piaget's cognitive theory and Gagné and Briggs' theory of instruction. These theories have provided new approaches in analyzing receiver cognitive development and pedagogical development. It would be extremely difficult to implement many designs in the field of communication technology without an understanding of the receiver and his/her environment. Once this is accomplished, there is a need to develop a proven system of communication, for without proper training the entire process becomes useless. Piaget's and Gagné and Briggs' theories represent the title "communication technology" well because one is communication-oriented and the other is technology-oriented. These theories not only are representative of the emerging field of communication technology, but also make a contribution that can be seen today. Their endurance supports the fact that Piaget's cognition theory and Gagné and Briggs' theory of instruction are continuing to be influential, yet these theorists are only part of the puzzle of aiding the communication process through the use of technology. Further investigation of media as a tool of communication has revealed insightful new information with the help of other theorists from the past, such as Gavriel Salomon. It is in understanding the theories of the past that the practice of the present can be conducted and the process of communication further explored.

Communication technology is a relatively young field, which appears to be expanding rapidly on many levels. Unfortunately, it is this expansion that is causing growing pains, thus hindering the exploration of new approaches in the field. It has become obvious that various experts believe new perspectives on past studies are required to break the frustrating hold of complacency and misinterpretation. In order for there to be advances in the field of communication technology, we must break the bonds of our self-imposed imprisonment. A revitalizing escape may be to turn to past

A symbol and a computer

researcher's concepts of media as tools of communication. The following section discusses the interaction among media, the message, and the receiver, as well as defines and discusses the use of media. The theorist that will be discussed is Gavriel Salomon with observations from several other prominent researchers. Gavriel Salomon was selected because of his well-thought-out discussions on media functions in the communication process.

Content Versus Technology

Gavriel Salomon's theory analyzes four major components of media: symbol systems, messages, technologies of transmission, and the receiver. According to Salomon, "symbol systems are the most important in that they appear to have different effects on mediating activities of information extraction and processing" (Salomon, 1974, p. 405). His theory argues that each medium has its own inherent symbol system, which therefore affects the message. These symbol systems offer added information and expand the communication process. Salomon believes that because of these symbol systems one medium differs from another and should be analyzed singularly. He points out that past studies analyze media technologies as one entity instead of individual instruments with their own identities. It is for this reason he observes that symbol systems are more crucial to communication than the technology of transmission (Salomon, 1974, p. 385). If symbol systems are so important to communication, then what are symbols and how do they become systems?

The symbols that Salomon refers to are "any objects, movements, gestures, marks, events, models or pictures that can serve as extractable knowledge" (Salomon, 1979, p. 29). Symbols serve as coding elements with rules and regulations and can be arranged into schemes. A symbol scheme becomes a symbol system when it is connected to a field of reference (Salomon, 1979, p. 31). For example, letters of the alphabet represent one meaning in reading and another meaning as variables in mathematical equation.

It appears that symbol systems encompass a great deal of information. These data are transferred to receivers when the systems are shared through common knowledge or by learning a new system. This information delivered in the form of symbol systems becomes the message. There have been many debates on the definition of *message* in regard to media. Symbol systems are an important part of a message, but there are also other elements that add to the information.

Marshall McLuhan pondered that there is more to the message than content in his theory that "the medium is the message" (McLuhan, 1964, p. 13). His bold belief that it is not the content but, rather, the medium that is the message created a new perspective on the importance of media selection. McLuhan was concerned with how information is delivered, not what the information contains. McLuhan saw the type of technology used as equal to or more important than the content. In fact, he believed that the technology is the content. Although Salomon refutes McLuhan's stand on the importance of the medium, it seems that they are in partial agreement for at least one reason. If various media contain their own inherent symbol systems that make them unique, then each medium's message must be affected in differing ways. Thus, part of the medium is the message. At this point, the transmission of technologies becomes important for the very reason Salomon disputes it. Media utilize many symbol systems simultaneously, but each medium has a unique system that is a part of the technology. It is these symbol systems that are a large part of how receivers comprehend messages. For example, hearing the national anthem is a different experience than watching the American flag wave over the Capitol.

Salomon does stress the importance of interaction among symbol systems, messages, and each medium. The receiver may share a better understanding of a particular medium's symbol system than various other media. For example, one individual may observe a still photograph and fully understand its message, while another individual looking at the same photograph may not comprehend the entire meaning. Thus, the first person has a better understanding of the photograph's symbol system. The first viewer is described as field-independent, the second as field-dependent. These individuals are the receivers who represent the final piece in Salomon's complex theory.

Finally, in the intricate process of media selection for communication stands the receiver. The question of the receiver's abilities, attitudes, and motivation is of utmost importance. These communicators have different levels of competency and varied goals. Because of these differing levels, it is difficult to match one medium with a group of receivers. Each receiver is not affected in the same manner by each medium. Hence, there is no superior medium, because the media are useful to different individuals, rather than indiscriminately restricted to one medium per group. Thus, knowledge of the receiver plays an important role in the interaction between the medium and the receiver. Prominent theorists, M.L. Koran, R.E. Snow, and F.J. McDonald discovered that people who are less field-dependent benefit from visual representations and that people who are more field-dependent gain knowledge from written material. This observation is a valuable piece of information in understanding the proper connection between the medium and the receiver. Therefore, it is evident that, because the receiver is difficult to define strictly, communication should encompass many elements to secure the transference of information and the acquisition of knowledge. If the interaction among media, cognition, and communication is so important, then symbol systems alone may not hold the key to acquiring knowledge. Media must also be rigorously defined and explored as much as symbol systems, cognition, and communication. Perhaps the uniqueness of the technologies' symbol systems is connected with the application of the medium.

The analysis of these theories represents an evolution from a communication event to the sequences of mediated messages. There is more to communication technology than what kind of media is to be selected. It is the audience that should be the focus of this field. Further, perfecting how the audience is going to respond through the aid of technology becomes the media producer's goal. The utilization of technology in communication requires a basis in sound theoretical practice, which has been richly provided by researchers. The study of these various theories gives insight into better ways of using technology in communication, as well as informs the producer on many nuances of the emerging field of communication technology.

The implications of this research are that the producer's focus is no longer on media selection alone but must include a thorough understanding of the audience and their constructed knowledge. A shift needs to occur in the field of communication technology from which medium is the best tool for communicating to which medium meets the needs of each communication situation. It is this paradigm shift that is having the greatest impact on the field of communication. As the use of new technologies continues to grow in society, the need for using this technology must be justified through research grounded in proven theories. Otherwise, we will continue to be mired in the endless debate that one "medium" is best and "that" medium will continue to change as technology evolves. Why reinvent theories when the road for using various technologies has been laid out for present and future media producers? The time has come to build bridges from past roads to new areas of communication.

The theories discussed all point to the importance of the audience and what each audience brings to the communication event. The process of communicating is complicated but manageable when discussed and analyzed in order to provide a variety of applications to each communication event. For this reason, it is important to study theories such as Piaget's theory of cognition and Gagné and Briggs' theory of instruction, as well as Salomon's theories. It is the process of communicating that is the key to aiding receivers in their quest for knowledge, and it is the producer's duty to make that process easier through the use of tools. Metacommunication, thinking about the communication process, is the shift the field of communication technology has made and needs to continue to make in order for this endeavor to be successful. Media producers need to keep these various theories in mind when designing mediated messages for maximum impact. The technology selected is a vital part of the message, but so is the audience. You can have the most elaborate and up-to-date Web page on the planet, but, if no one visits it, was there ever a message at all? It is the same as the old saying "If a tree falls in the forest and no one is there to hear it fall, did it make a sound?" The answer for a media producer is that, without an audience, communication cannot take place, and the audience is the key to designing successful mediated messages.

A person at a computer

INFORMATION IN THE TWENTY-FIRST CENTURY

Researchers have helped lead the way into the twenty-first century and modeled how technology should be used for successful mediated communication. Their systematic approach to constructing a message for a particular audience has a proven track record in education and can be adapted in today's uses of communication technology. We do not have to reinvent the wheel with each new technology but merely apply sound theories to the practical design of mediated messages. Successful media producers will keep in mind that each technology has its own inherent symbol system, which may work better with certain audiences, such as video rather than textual information for receivers from "Generation X." The Internet or audio might be a better medium selection for those of "Generation D," depending on the situation. The medium selection should depend on several factors besides media symbol systems. The receiver's background and experience need to be considered, as well as the actual message. Using technology to communicate should never be a random choice but, rather, a well-thought-out, systematically constructed process. Why do advertisers use Britney Spears in radio and television commercials to sell Pepsi to teenagers?

A pie in pieces illustration

UT
NIU
BGSU
DePaul

The Whole Is Greater Than the Sum of the Parts

The communication process begins with media producers becoming critical viewers of mediated messages. "Media influence exists to the extent that media creators [senders] and audience [receivers] uncritically assemble and accept these symbol systems as standards for evaluating and making comparisons, for interpreting themselves and their world" (Brown, 1991, p. 27). To understand how to construct successful mediated messages, media producers must understand the parts that make up the message, because in comprehending the parts the thrust of the whole message becomes clear. Thus, media producers need to become critical viewers of mediated messages. It is only by decoding the various constructed meanings within media that a producer can learn if successful communication is taking place. A critical analysis of mediated messages can shed new light on social, cultural, and political issues projected on or reflected within the message of the media by revealing how meaning is constructed in that message. It is beneficial to producers, business trainers, and communication technology professionals, because knowledge of how meaning is constructed helps them to understand better the viewers of this material, thus aiding in the creation of better designed media that are more focused on the viewer and how the viewer gleans information.

A person watching a TV

The Rise of Critical Viewing

The rise of interest in critical viewing, decoding meaning in mediated messages, issues can be traced back to critical viewing projects that were conducted in the late 1970s, when a need for these projects became apparent. As Fred Silverman (1986) observed, most Americans spent so much time in front of the television; it had become important to design and develop critical viewing skills in schools. Therefore, four projects were developed at the elementary, middle school, high school, and higher education/adult levels. According to Frank Withrow (1980), the Southwest Educational Development Laboratory (SEDL) developed the elementary project;

a public television station in New York, WNET, developed the middle school project; the Far West Laboratory in San Francisco developed the high school project; and Boston University's School of Public Communications developed the adult as well as the higher education project. The purpose of these projects was to create critical viewers. The results were a set of criteria to develop critical viewers at the various levels. The U.S. Office of Education over a two-year period supported projects. There were other projects, conducted by the Idaho Department of Education, the American Broadcasting Company, and the National Education Association. The basis of these projects was similar to that of the U.S. Office of Education projects, established to develop criteria for making children and adults critical viewers. However, most of the criteria developed were aimed at children.

A TV

Critical Viewing Skills

The criteria developed by the SEDL for the elementary level was material, on how to use television to complement language arts, science, and math, such as cue cards, which guided teachers and parents (Withrow, 1980, p. 56). The criteria developed by WNET for middle schools was a book of worksheets to help students improve their vocabulary and increase language and reading arts skills (Withrow, 1980, p. 56). The criteria developed by Far West Laboratory in San Francisco for the high school level was designed to teach students four skills: to evaluate and manage one's own TV viewing behavior; to question the reality of TV programs; to recognize the arguments used on TV and to counterargue; and to recognize the effects of TV on one's own life (Withrow, 1980, p. 56). The criteria developed by Boston University's School of Public Communications for the higher education/adult level involved four major subject areas: television literacy, persuasive programming, entertainment programming, and informational programming. The goal was to familiarize adults with the medium in order to aid them in making critical judgments. According to Bradley Greenberg and Carrie Heeter (1983), "a major effort funded by the U.S. Office of Education (USOE) focused on creating curricula and related materials to aid development of critical viewing skills" (p. 48). These projects and the others discussed earlier all focused on developing critical viewing skills but did not measure the effectiveness of the skills.

Most studies conducted on critical viewing skills model critical viewing skills with limited information describing the effectiveness of individual skills. For example, a study conducted by John Splaine (1988) on the 1988 presidential election models seven critical viewing skills. These skills are very similar to the critical skills mentioned earlier. They basically consist of the understanding that television is manipulated technically and by subject matter choice. The results of the critical viewing studies are measured by the improved performances of the students participating in the projects. Although exact figures are not given, most projects report improved performances from students involved in using critical viewing skills. For example, a study on the effects of critical viewing skills "revealed significant knowledge gain by students involved in the curriculum" (Watkins, Sprafkin, & Gadow, 1988, p. 165). Thus, media producers who practice critical viewing skills should also improve their knowledge.

According to James Anderson (1980), although these educational projects popularized critical viewing, "the notion that critical skills direct the processing of information is an old one" (p. 64). For example, critical reading skills and listening skills have been a part of educational programs for years. Students are taught to critically read classical literature, and "students are placed in social studies classes where they supposedly learn enough about our political social structure to perform as responsible voters" (Robinson, 1974, p. 3). It was inevitable that critical skills were addressed to media; thus, critical viewing skills were born.

The Producer and the Mediated Message

As critical viewing has developed, who the producer is and what the mediated message is have become the vital connection between critical viewing skills and critical viewing analysis. According to Len Masterman,

> A critical understanding of media, then, will involve a reversal of the process through which a medium selects and edits material into a polished, continuous and seamless flow. (1985, p. 127)

Critical viewing analysis gives the producer the ability to analyze media messages by breaking the whole into parts to reveal the multiple meanings constructed within any mediated message. As Belland, Duncan, and Deckman (1991) observed, "whether the objects of criticism be scientific findings, aesthetic works, or technological gadgets or processes, sound criticism is the exercise of intimate, informed judgment" (p. 153). It is this act of judgment in which all producers participate. Although critical media analysis has been an accepted practice in other fields for years, it is fairly new to the field of communication technology. For this reason, the acceptance of this form of analysis in this field has been problematic. According to Gary Shank (1990), these problems stem from the quantitative grounding of the field. Nonetheless, studies are being conducted utilizing this type of analysis, and information is being uncovered.

Critical viewing analysis is a thorough tool, which allows the producer to understand the "organizing ideas by which people develop perspectives about their relationship to the world" (Himmelstein, 1981, p. 97). It is the media producer's job to create, decipher, and interpret the media message. This aids the producer and the viewer in understanding the full meaning and effect of the mediated message on the viewer. Critical viewing analysis is akin to critical thinking, which is "a way of looking at the world" (Newcomb, 1981, p. 12). The producer becomes an important explorer of the message, and the message becomes the understood instead of the experienced. Still, the experience is important as to how meaning is constructed by the viewer. According to Brown,

> The producer, and the critical viewer, must be grounded in two essential areas of this act of critical judgment, the area of facts (what is informational data) and the area of norms or standards (what ought to be criteria). (1991, p. 24)

This deconstruction of mediated messages helps the media producer become a critical viewer. The importance of this transformation from viewer to critical viewer cannot be overstated, because the viewer becomes an active participant in understanding the mediated message from several positions. In other words, the viewer is no longer a passive sponge but an active part of the media experience. This type of viewing can be defined as "an awareness of how symbols become meaningful and/or almost any activity that distances viewers from the text" (Cohen, 1994, pp. 101–102). This repositioning of the viewer allows the viewer to comprehend the many layers of meaning in any one media message. This repositioning helps the producer better construct the message through a systematic process. Heller described this process as "the viewer assumes that meaning resides within three spheres: (1) The work or event; (2) the person examining the work; as well as (3) the time and place in which the person finds himself/herself" (1982, p. 848). It is this interaction that produces the meaning in a mediated message and allows the producer to place messages unconsciously gleaned from an initial viewing.

The average spectator of a mediated message views the information and accepts it as a natural representation of society. As Heller stated, "the audience is not a passive receptor of mass persuasion" (1982, p. 850) but, rather, an active acceptor of what appears to be natural, while the critical viewer breaks the mediated message down into pieces that can be identified and more fully understood. The study of this interaction among the mediated message, the producer, and the environment in which the mediated message is viewed explains the tenuous relationship between the producer and the media message. It is a never-ending quest to explore and describe "who says what to whom." In other words, who is producing and controlling the meaning in any media text? The answer to this question aids the producer in deciphering how viewers are participating in the process. This process creates a viewer's reality by using "a reactionary mode of representation that promotes and naturalizes the dominant ideology" (Fiske, 1987, p. 36). The producer and the mediated message have a delicate relationship, in which the producer strives to understand the world better by exploring the mediated message. It is in analyzing the message that the producer finds meaning within and outside the message. This tenuous bond can reveal information not easily gleaned from mediated communication on the initial experience. A producer needs to analyze a media message on various levels and at least several times to conduct a thorough investigation.

It is through critical viewing analysis of media messages that one can come to understand the power of the media and who controls them. This type of inquiry guides the media producer to a new level of comprehension and, thus, empowers him/her. We strive to understand our world, which is extremely symbolic, a world that uses symbols to inform and communicate through various media. For example, a particularly powerful medium consists of Web sites, and a critical viewing analysis of this medium could reveal information to the viewers and designers.

Critical Viewing Analysis Framework

Most critical analysis studies discuss the importance of becoming a critical viewer of media messages. They may utilize different methods to analyze a message, but the intent is the same. These critical viewing researchers have been working toward a more media-literate society by uncovering the multiple layers of meaning in mediated messages. It is through this process that producers can become more knowledgeable about how meanings are produced. The importance of creating critical viewers is evident; most, if not all, of the information available today is mediated. Hence, any analysis that critically views these mediated messages adds to the knowledge base on how to perform critical viewing analysis and prepares producers to design better mediated messages. It is imperative that the practice of critically analyzing media messages continues to be performed to break the cycle of the routine acceptance of poorly designed messages.

"Effective viewing of real world images requires a critical sense for both information and shades of meaning" (Adams & Hamm, 1988, p. 82). It is essential that producers of mediated messages learn to sort out the meaningful information from the trivial by building a critical viewing analysis framework. According to John Long (1989), this framework should consist of intervention, goal attainment, cultural understanding, and literacy (p. 13). Intervention allows critical viewers to determine fact from fantasy, while goal attainment allows critical viewers to monitor and understand their own reasons for attending to the electronic medium. Long also observed that cultural understanding is a matter of the critical viewer's understanding where in the social schema media exist. A combination of these three factors [intervention, goal attainment, cultural understanding] creates a literate viewer. Heller agreed with Long but was more general in her observation that "description, interpretation and evaluation of work are dependent upon the interaction of all three spheres" (1982, p. 848). These researchers see critical viewing analysis as a relationship among the message, the producer, and society. Furthermore, Long believed the last vital component of critical viewing is an understanding of semiotics (symbols) which suggests that the interaction of message, producer, and society involves the construction of meaning through symbols.

Neil Postman (1985) observed that, around the turn of the twenty-first century, Western society left print-typographical culture behind and entered a new "age of entertainment" centered on a culture of the image. Furthermore, Kirrone agreed with Postman and stated, "We are becoming less involved with print media and more involved with images—photos in magazines, videotapes, movies and so on—and these images are altering how we think and learn" (1992, p. 60). These mediated messages are guiding viewers and affecting their knowledge without their ever realizing the process. As Kellner pointed out, "critical literacy in a postmodern image culture requires learning how to read images critically and how to unpack the relations between images, texts, social trends and products in commercial culture" (1988, p. 43).

Communication technology is very much a part of the commercial culture and in need of critical analysis. Mediated messages are cultural texts, and it is through analyzing these cultural texts that people come to understand the world. Media try to guide the viewer's behaviors and/or attitudes to meet specific objectives. The process the media producers use to guide the viewers is an important link to the type of behavior/attitude modification that may or may not take place. It is important to understand that not all behavior/attitude modification is negative but can be a positive improvement for the viewer. Critical viewing analysis of media messages makes it possible for the critical producer to view how audience members see themselves or want to see themselves.

Critical Viewing Model

Critical viewing refers to a close reading of a mediated message, one that focuses on the signification of a message. Critical viewing of a mediated message is very different from casual viewing, because it takes into account a strategy that informs the producer of what may be present in a message and necessitates a method to describe how one systematically designs a message. Media producers can follow a model that uniformly encodes the meanings in mediated messages. The producer identifies connotation, denotation, ideology, and literacy based on the culture created by the media message. However, instead of just looking at the pattern emerging from the columns of information, this model goes one step further by involving the four areas of critical viewing.

The model is a four-step process, in which each step designs the message from two perspectives. In step one, the producer identifies connotation (such as photograph of the U.S. flag) but then analyzes intervention. Intervention involves understanding that the process of perception may be altered through self-regulatory skills. You see the flag and what comes to mind depends on your belief system. This is the most important step in the process, because it is at this level the viewer stops being a passive viewer and becomes an active viewer. This is accomplished by the viewer's understanding his/her internal perceptions and biases. Every individual brings a set of beliefs to every situation. Discovering how to separate one's own perception from the media message is the key to becoming a critical viewer. Step two of the model consists of identifying denotation (represents the United States of America) and looking at goal attainment. Goal attainment entails monitoring and understanding one's own reason for viewing the media message. There are two types of goals: personal and others (producers of the message). In other words, the viewers and the media message each has a set of goals. It is the responsibility of the critical viewer to understand his/her personal goals (WWII veteran) and the goals of the media message (selling life insurance). The third step of this critical viewing model is identifying the ideology of the codes and cultural understanding. Cultural understanding is knowing where in the social schema the media message exists. It is the opposite of intervention, because the viewer works on the external awareness of others' perceptions and biases. The viewer is concerned with understanding the media message being viewed through perceptions and biases. In other words, what assumptions does the mediated message create about the characters, environment, and so on? The assumption might be that World War II veterans will respond to a media message selling life insurance with the U.S. flag in it. The final step in this model is literacy. Literacy is comprehending the overall grammar of the mediated message. It is at this step in the process that the producer can take all the knowledge gained from the first three steps of the critical viewing model and interpret the overall meaning created from the mediated message. See Fig. 2-1. By thoroughly following steps one through three in the model, a producer should naturally become a literate viewer, making the final step in the process the result of successfully applying the model.

This model not only encodes the signs that create meaning in a media message but also finds and distinguishes the viewer of that message giving a richer grasp of the construction meaning. The producer can systematically encode a media message,

FIGURE 2-1
A critical viewing model

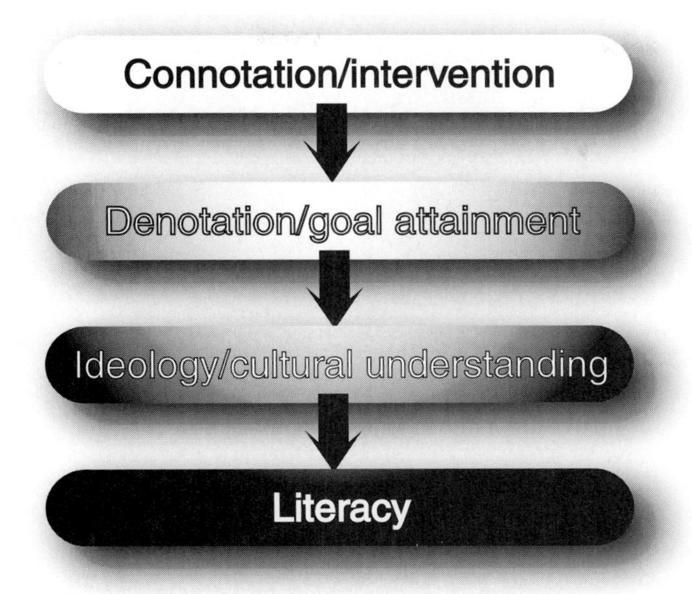

creating meaning and developing successful communication. This model creates skilled producers who are trained and prepared for authentic assessment and focused design. In turn, skilled producers develop more valid and reliable designs. The model fills in the valuable information about mediated messages that for so long has been elusive. The result is the birth of critical viewing producers. According to Martha S. Feldman (1995), "the primary aim is to be able to make sense out of phenomena that were previously puzzling or to be able to make new sense out of phenomena that were not previously fully examined" (p. 30). This model aims to make sense out of how meaning is constructed in media and is defined by an audience. In the end, the whole message is truly greater than the sum of the parts of a mediated message.

LESS IS MORE

This chapter discussed many theories and researchers, as well as the various ways a media producer can successfully craft media messages using these proven theories. The discussion has led to the importance of knowing your audience and systematically designing your message using the correct media to ensure success. The success begins and ends with the audience and how you use the various available technologies. Whatever the technology selected, a focused, streamlined approach will engage the audience every time. The old adage "less is more" speaks volumes to an audience and should be every media producer's mantra. In addition, always remember the combination of audience, message, and technology and do not let one element dominant the other two, or you will create junk communication. The following chapters will attempt to explain the practice of the successful design of various media (print, audio, video, digital, and Internet) and to help you avoid the very common pitfalls of junk communication.

Simple design versus busy design

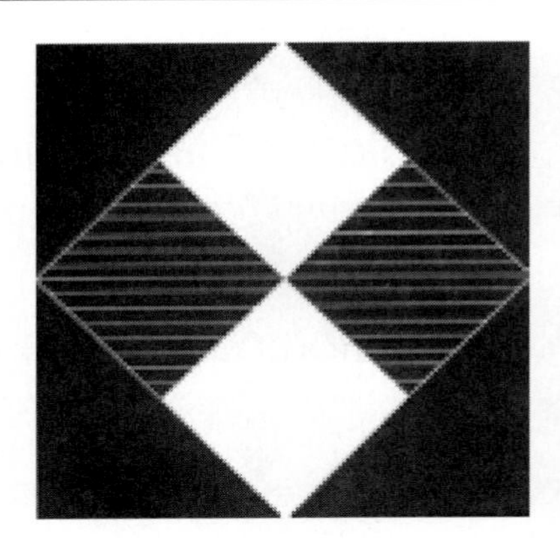

BIBLIOGRAPHY AND SUGGESTED READING

Adams, D., and M. Hamm, (1988, March-April). Video Technology and Rural Development. *The Social Studies*, 79 (2), 81–83.

Anderson, J. (1980). The Theoretical Lineage of Cirtical Viewing Curricula. *Journal of Communications*, 30 (3), 64–70.

Belland, J.C., J.K. Duncan, and M. Deckman, (1991). Criticism as Methodology for Research in Educational Technology. In D. Hlynka and J.C. Belland (Eds.), *Paradigms Regained: The Uses of Illuminative Semiotic and Post-modern Criticism as Modes of Inquiry in Educational Technology*. (pp. 151–164). N.J. Englewood Cliffs, Educational Technology Publications.

Brown, J. (1991). *Television Critical Viewing Skills Education: Major Literacy Projects in the United States and Selected Countries*. Hillsdale, NJ Lawrence Erlbaumcnel Associates, Publishers.

Cohen, J. (1994). Critical Viewing and Participatory Democracy. *Journal of Communications*, 44 (4), 98–113.

Feldmar, M.S. (1995). *Strategies for Interpreting Qualitative Data*. Thousand Oaks: Sage Publications.

Fiske, J. (1987). *Television Culture*. London: Methuen and Company.

Gagné, R.M., and L.J. Briggs. (1974). *Principles of Instructional Design*. New York: Holt, Rinehart and Winston.

Gagné, R.M. (1977). Instructional Programs. In M. Marx and M. Bunch (Eds.), *Fundamentals and Applications of Learning* (pp. 404–428). New York: Macmillan.

Gagné, R.M. (1986). Instructional Technology: The Research Field. *Journal of Instructional Development*, 8(3), 7–14.

Greenberg, B. and Heeter, C. (1983). Television and Social Stereotypes. *Prevention in Human Services*, 2(1–2), 37–51.

Heller, M. (1982). Semiology: A Context for Television Criticism. *Journal of Broadcasting*, 26 (4), 847–854.

Himmelstein, H. (1981). *On the small screen: New Approaches in Television and Video Criticism*. New York: Praeger.

Kellner, D. (1988). Reading Images Critically. Toward a Postmodern Pedagogy. *Journal of Education*, 170 (3). 31–52.

Kirrone, D. (1992, September). Visual Learning. *Training and Development*, 46, 58–63.

Koran, M., R. Snow, and F. McDonald. (1971). Teacher Aptitude and Observational Learning of a Teaching Skill. *Journal of Educational Psychology*, 62, 219–228.

Long, J. (1989). Critical Viewing and Decision Strategies (Doctoral Dissertation The University of Utah, 1989). *Dissertation Abstracts International*. 50, 1471.

Masterman, L. (1985). *Teaching the Media*. London: Camealton Publishing Group.
McLuhan, M. (1964). *Understanding Media*. New York: McGraw-Hill.
Newcomb, H. (1981). Television as Popular Culture: Towards a Criitical Based Curriculum. In M. Plaghaft and J. A. Andersan (Eds.), *Education for the Television Age*. (pp. 9–18).
Piaget, J. (1954). *The Construction of Reality in the Child*. New York: Basic Books.
Piaget, J. (1957). *Logic and Psychology*. New York: Basic Books.
Postman, N. (1985). *Amusing Ourselves to Death*. New York: Viking.
Robinson, D. (1974). Film Analyticity: Variations in Viewer Orientation (Doctor al Dissertation, University of Oregon, 1974). *Dissertation Abstracts Invernational*, 44(4), 98–113.
Saettler, P. (1990). *The Evolution of American Educational Technology*. Englewood, CO: Libraries Unlimited.
Salomon, G. (1974). What Is Learned and How It Is Taught: The Interaction between Media, Message, Task, and Learner. In D.R. Olson (Ed.), *Media and Symbols: The Forms of Expression, Communication, and Education. 73rd Yearbook of the National Society for the Study of Education* (pp. 383–406). Chicago: University of Chicago Press.
Salomon, G. (1979). *Interaction of Media, Cognition, and Learning: An Exploration of How Symbolic Forms Cultivate Mental Skills and Affect Knowledge Acquisition*. San Francisco: Jossey-Bass.
Shank, G. (1990). Qualitative Versus Quantitative Research. A Semiotic Non-problem. In T. Preuitt J. Deel and K. Haurth (Eds.). *Semiotics* (1989) (pp. 264–270). Washington, DC: University Press of America.
Silverman, F. (1986, March). Advances in Technology Translate into New Opporunities. NASSP-*Bulletin*, 70 (488), 50–56.
Splaine, J. (1988). Televised Politics and the 1988 Presidential Election: a Critical View. *Georgia Social Science Journal*. 19 (2), 7.
Watkins, L.T., J. Sprafkin, and K.D. Gadow, (1988). Effects of a Critical Viewing Skills Curriculum on Elementary School Childrens Knowledge and Attitudes about Television. *Jounral of Educational Researcher*, 81 (3), 165–170.
Withrow, F. B. (1980, September-October). Why We Need Critical Viewing Skills. *Today's Education: The Journal of the National Education Association*, 69 (3), 55–56.

3 Perception

How to "see" using technology

Studying visual communication history helps us develop a language of visual awareness. Determining whether a design solution is good or bad relies on the ability to judge effectiveness. These judgments defining the success or failure of a graphic communication are irrelevant in an instance lacking intent by design. Any attempt at solving a visual communication problem can be considered successful merely because an outcome has been produced—good or bad. Historically, the value of this outcome was often arbitrarily subjective. Inevitably, an audience experiencing this solution could read the visual message in possibly unintended ways. Ultimately, the success or failure of a graphic communication is embedded in the audience's accurate perception of what the designer intended. Until well into the second technology trend in visual communication, creators of drawn and printed material were unconcerned with all but the most basic of intents—to get the message out. These early media communicators used the reigning technology, whether it was a stylus or block print, most often for the sake of the ability to print alone. Bibles, governmental rules, and inventories of a kingdom's resources had to be printed and distributed. This kept most of the reading public minimally informed. The ability to read and comprehend what was printed became very important as the technologies in printing improved. The movable type printing press drove the education of audiences. A recurring problem surfaced in the visual communication process. A message was conveyed to the audience of a graphic communication regardless of whether that audience could read the information as intended or not. The skills involved in being able to direct that perception lie in the realm of the graphic communicator. Therefore, the necessity for graphic design communicators to develop visual literacy through the technologies of their media became vital to effective visual communication. See Fig. 3-1.

The technology trends form the visual design for that era. For example, while the Middle Age monks were creating illuminated manuscripts of the Bible, very few people actually were privy to viewing, let alone reading, these books. The technology of handcrafted books was inadequate for mass consumption. It was simply too expensive and slow. Most scholars agree that the two most important communication technologies in recorded history are the movable type printing press and the personal computer. See Fig. 3-2. Why are these technologies so important? In order to understand their impact, one must first appreciate the power of these tools. The printing press changed the way people received and interpreted text information. The mass production of books made

FIGURE 3-1

Three main events in visual communication technology

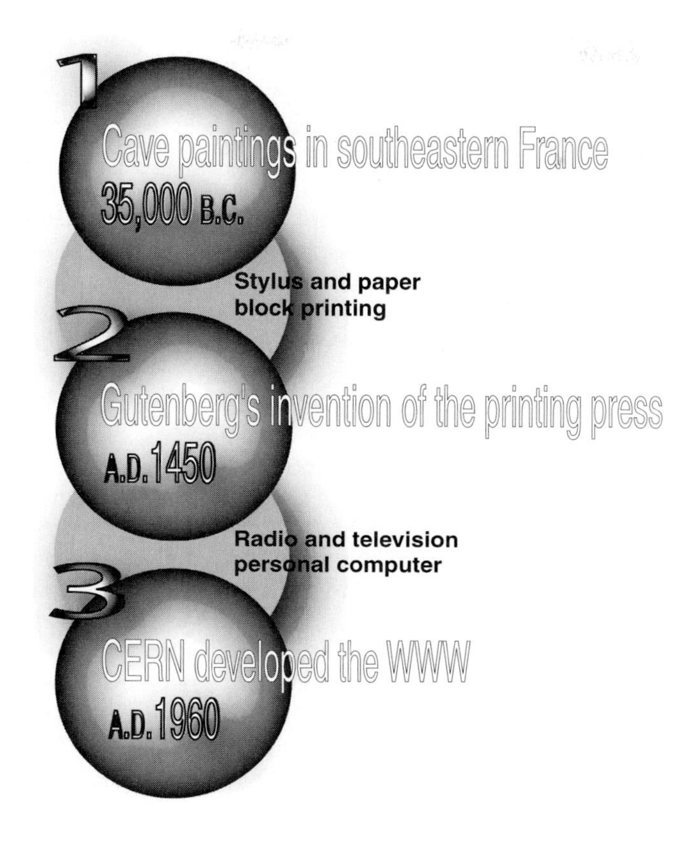

FIGURE 3-2

Gutenberg's printing press and the classic Macintosh personal computer

Landmark Technologies in Visual Communication

possible by the printing press paved the path for religious, public, and private education. A reading public changed the way humans govern themselves. The impact of this visual communication technology is virtually undeniable. At the same time, creators of printed media were beginning to understand that a dichotomy existed in the visual communication process. The intended message was inextricably tied to the method in which it was communicated, while the content of the message was directly influenced by its visual organization. A visual communicator needed to understand the impact of the technology in which the form of the message was produced and, at the same time, be aware of the psychological perceptions created in the minds of the audience by the visual organization of the elements on the page.

Meanwhile, as the masses were consuming printed media, experiments in new, "electronic" ways to communicate were being explored. Movies and television pushed visual communication to unimagined perspectives. The audience's perception of a message was manipulated through image genres and special effects. The broadcast of these visuals to every corner of the planet became commonplace. Printed communication and electronic communication lived in harmony; these media complemented each other—one promoting the other. Producers and designers became increasingly aware of the power in their media technology, especially in the past fifty years. Rapid technological advances in printing and electronic media taught the public to develop a sort of cynicism concerning visuals. The mass audience's perception of media technology seemed to be "You can't believe anything you read in the newspapers or what you see on TV." This perception was promoted by the ever improving technical skills of the artisans who created special effects, animations, and graphic design.

As versatile as printing has become, it still takes a relatively long time to be created and distributed. In addition, printing in color, on anything other than plain paper, and mailing printed pieces is expensive. Getting the message out turns into an extremely complicated process involving marketers, designers, printers, distributors, and managers. TV bears some of the same burdens, as well as its own unique set of limitations. TV communicates most often to a single family or an individual. Rather than settling with the limitations of these technologies, creative communicators began innovations in the merging of these media. The invention of the personal computer provided a new avenue for visual communicators. Using these media technologies in cooperation with each other creates a hybrid integration of the simple "sender to receiver" model. Audiences can experience messages rather than merely view them. Visual communicators are now mandated to create messages that engage the viewer both mentally and physically. Designers can fashion graphic communications with a virtually unlimited palette of media stimuli. Armed with these tools, designers are using media technology to help the audience "see" with all their senses while coming away with the intended message. The future challenge for these visual communicators is to balance hybrid media to create the most intentionally effective message. What does one want to say? Which media should be utilized to say it? Whatever the message, it must be delivered through the media technology of the time.

VISUAL ORGANIZATION

Meeting the challenge of delivering visual communication through various media technology is the first step in the design process. The second is visually organizing graphic elements so that they make sense to the audience. The process of successfully organizing elements is identical on a page, the TV screen, or a monitor, yet how is successful visual organization achieved? The theories and practices developed during the Bauhaus period are the most prevalent methods used to explain predictable outcomes for a design problem. This period in history is often referred to as the birth of formal graphic design theory. Thousands of very intelligent and practiced design professionals have offered a plethora of interpretations of these theories. Most of these theories are an explanation of basic design principles, which when applied successfully produce the intended message to a target audience. The main category of concern for a graphic communicator is the Gestalt psychology theories that include the consideration of visual perception, balance, and figure/grand grouping.

The principles of Gestalt theory depend on the human behaviors of "pattern seeking." In other words, people attempt to organize space both mentally and physically for psychological contentment. The whole is greater than the sum of its parts. Individual elements can be analyzed and identified singularly, but it is the synthesizing of these parts that gives full meaning to what is seen. For example, the human body can be identified as thousands of interactive parts: head, arms, torso, legs, and so on, however, considered as a "whole," that body becomes a human being, a personality, a sister, a dad, a professional, and so on. Similarly, the organization of parts in a visual frame of reference can be identified as individual elements: text, photographs, lines, shapes, illustrations, and so on; however, considered as a "whole," that frame becomes an advertisement for soda or a film about lost love. Furthermore, this perceptual psychology teaches designers and producers that there are predictable reactions by most audiences to intentional Gestalt patterns of visual organization. With this knowledge, communicators can actually foresee certain responses by their target audience as a result of exposure to specific visual stimuli. Early historical attempts at evaluating the effectiveness of visual communication consisted of subjective judgments as to whether a design was pleasing to an audience. Gestalt perceptual psychology provides a concrete visual measuring tool that helps define good and bad design. Graphic designers who understand and use these principles are able to discern objectively between effective and ineffective design solutions. Through these visual perceptions, a graphic communicator is challenged to balance the form a design takes with the importance of the content that must be delivered. The actual motivational influence of arranging and manipulating graphic elements over the reactions of a viewer is measurable. Many studies have been conducted since the turn of the twentieth century to prove these standard reactions based on human physiology. Our discussion of Gestalt perceptual psychological theory is limited to the awareness of human responses to visual stimuli. Students of visual literacy will find that the application of simple Gestalt theories to graphic problems is more than adequate for developing meaningful design solutions. Applying these principles in balance with each other creates the nuances required for an effective aesthetic.

Visual Perception

Visual perception is the acceptance by an audience of a graphic message through the processes of seeing. A viewer's perception of any visual stimuli is based on one's previous experiences, expectations based on those experiences, and ultimately the physiology of human brain structure. Interestingly, these cues of common human experiences are the hooks used by emerging media technology experts to produce unexpected results to a visual. If one truly believes that humans can travel via a transporter, fly without machinery, or meet with extraterrestrials, then that person has surely gone crazy, yet, in the realm of movies, TV, and graphics, these instances are made totally believable by the skillful manipulation of one's perceptions. Human experience as an earthbound being is the prevailing perception of our reality, but, when one is viewing a science fiction effect produced by media technologists, that perception is altered. One enters into an environment where expectations are based on unreality.

As compelling as special effects are in the movies, the real meat of altering perception is in the area of persuasive content. The Star Trek scenario describes how a sophisticated creative technician is able to skew our perception of reality. The skewing of experience in this sense lasts only as long as one is emerged in this visual unreality. On the other hand, graphic communication that manipulates reactions without conscious consent is where subliminal messages can be transferred to an unsuspecting audience. The enhanced ability to embed messages into visuals causes ethical and moral considerations involved in what could be described as "subconscious seduction." An audience's perception of a visual communication may be experienced at many psychological levels, and these multifaceted perspectives often create mixed messages. The intended messages of designed communication belie the competencies of its producers.

The intent of the designer is paramount when developing a visual communication. If the intent is to please oneself, then most often the communicator is a fine artist. At this level, the visual communication is purely personal, and an audience's pleasure or displeasure is secondary. If the producer's intent is to please a client, then the visual communicator steps into the shoes of a graphic designer. Graphic designers attempt to persuade, to inform, or to inspire through visual communication. Some examples are an advertising designer persuading an audience to buy a product, a public relations designer informing an audience about an upcoming event, and a graphic designer inspiring an audience to facilitate change through a poignant poster. The content of these intentions is manifested in the form of a visual communication through text, graphics, photographs, and moving imagery. As the graphic designer or producer lays out the elements of a visual experience, he/she bases visual organization on human behavioral tendencies and attitudes. The expectation that people relate visual elements to previous experience can often be predicted and exploited by an educated graphic communicator.

Balance

Balance is the key to successful graphic communication. Balance is the delicate weighing of elements to produce a psychologically stable visual environment. Balance can be achieved visually by creating the perception of weight and direction. Concepts related to visual communication balance through extremes: chaos is balanced by order, figure is balanced by background, content is balanced by form, and creativity is balanced by focus. Using the opposite visual force to create perceptual weight and direction allows the designer to achieve harmony. To realize visual harmony, one needs to understand the basic human tendency to organize space. In a painting by Jackson Pollack, an abstract expressionist artist of the 1940s and 1950s, the graphic elements are initially confusing. Pollack's style was called "action painting," and he literally threw his paint in long streams of color onto enormous canvases. Even though Pollack had repeatedly stated that he had no intention of creating any recognizable imagery, audiences continually reported seeing images in his work. The human tendency toward pattern seeking compelled viewers of Pollack's paintings to organize his totally abstract gestural content into comfortable, recognizable elements: faces, bodies, trees, and the like. The familiar objects detected in the paintings were the result of arranging a chaotic abstraction into an ordered visual. See Color Plate 3-1 in color insert.

This tendency toward order can be extended to both concrete and abstract concepts. Why do people close the door when walking out of a room or straighten a crooked picture? Why do we need to end a relationship or finish a project? These are attempts at creating order through the principle of closure. It can be applied to graphic communication in a viewer's mental tendency to finish or close shapes. For example, it is only necessary to show the top 50 percent of text for most people to read information. See Color Plate 3-2 in color insert. By applying the principles of human experience and closure, the audience of this particular visual will mentally complete the rest of the letters. By further exploiting the principle of closure, a graphic designer can engage previously studied responses by an audience. When audiences project their emotional reactions into a visual communication, they are responding according to what visual perception psychologists call isomorphic correspondence. For example, an image of a car crash reminds the viewer of human injury or pain, a photograph of a juicy hamburger at lunchtime triggers a feeling of hunger, and a commercial showing a luxury car driving through beautiful terrain promotes envy pangs. The principle of correspondence allows a graphic designer to lead the target audience to an expected reaction, thereby persuading them to act accordingly. Extending an audience's responses to a visual beyond the physical limitations of the graphic is employing the principle of continuation or what may also be defined as kinesthetic projection. A perfect example of this reaction is the ability to lead viewers in a desired direction simply by exposing them to a printed arrow. Surely here one can see the evidence of the whole being greater than the sum of its parts. An arrow pointing in a certain direction conveys so much more than the graphic itself. It says, "This way, because there is something in this direction that must be visited." It says, "Do not go that way because it is not the right direction." The message gleaned from a single arrow graphic is relatively enormous in proportion to the graphic itself. Closure, isomorphic correspondence, continuation, and kinesthetic projection allow a graphic communicator to rely on tendencies in studied human behavior to lead an audience to finish, close, complete, imagine, or follow.

Balance can also be explored by looking at the perceived visual weight of graphic elements within a frame. One way of using visual weight in a composition is through symmetrical balance. Symmetrical balance is based on identical elements reflected equally in a frame of reference. Balance occurs through the similarity of elements. The Ancient Greeks and Romans perfected symmetrical balance in their architecture as well as their art. The Greeks developed a "golden section," dependent on mathematical divisions of a simple rectangle. This precise arrangement utilizing mathematically accurate formulas became known as classical form in design. Symmetry was the cornerstone of this style. Symmetrical balance can be easily identified by its repetition of identical points, lines, planes, and volumes. If one were to fold a perfectly symmetrical design in half and then quarters, when unfolded all the quadrants would display identical graphic units. The perceived visual weight of the figure versus background seems equal even if the ratio of positive to negative space is mathematically unequal. Symmetry does not always rely on balanced visual weight but on the likeness of graphic elements top to bottom and right and left on a frame of reference. The audience's response to symmetrical designs is a feeling of gravity and stability. These designs are classified as conservative, static, traditional, and sometimes dull. Symmetrical visual organization is utilized when a graphic designer wants to establish unmovable structure, comfort, and calmness. Symmetrical designs do not automatically produce the perception of balanced visual weight, but most often the nature of symmetry in the equal division of space produces an equal weight effect. Still, at some points, symmetrical designs can be perceived as too heavy or too light.

On the other end of the spectrum, designs based on asymmetry can produce balanced visual organization also. Asymmetry is based on differing elements creating balance through contrast. This type of balance is created by the contrasts between graphic elements, rather than by their similarities. During the art historical period defined as Art Nouveau (1900–1912), an artist named Henri Toulouse-Lautrec developed posters and paintings heavily reliant on asymmetrical balance. See Color Plate 3-3 in color insert. His work was visually exciting, intentionally unstable, and refreshingly dynamic. Lautrec's paintings were organized and manipulated to achieve tension between contrasting elements. Dividing space asymmetrically creates unexpected movement. In asymmetrical designs, the viewer perceives the resulting graphic as visually balanced due to the careful organization of weight relationships in what is considered positive spatial elements and negative spatial elements. Note that asymmetry is present simply by the existence of dissimilar, contrasting graphic elements in a frame of reference where these elements may or may not display balanced visual weight. To maintain balance, the success of asymmetrical design is in the attention paid to developing strongly evident contrast between elements. These contrasting elements carry commonly perceived directions and weights.

To avoid visually unbalanced designs, the graphic communicator must be aware of perceptual comparisons viewers tend to make when seeking patterns in a design. The ways to produce weight and direction are numerous, and the following categories are those that are commonly accepted as successful.

- Elements on the bottom of a frame seem "heavier" than elements at the top because of gravitational human experience.
- The center of a design allows for more visual weight than the edges around the frame.
- Black or dark, closed shapes are more often defined as positive elements and thereby as possessing more visual weight than lighter, open shapes.
- Most viewers see vertically and horizontally oriented elements as more stable weight, because these strict 90-degree and 180-degree angles reflect the nature of the horizon and human standing postures.
- More complicated shapes seem visually heavier than simpler shapes, such as squares and circles.
- A shape that is isolated in a frame of reference carries more visual weight than evenly spaced elements and draws the viewer's attention.
- Figures, and especially faces, carry more weight than abstract, nonrecognizable subjects do.

- In the Western Hemisphere, we are conditioned to rate the left side of the page as more important than the right side; therefore, eye movement from left to right seems easier.
- Because we are taught to read from left to right, directing viewers in the opposite directional flow tends to be more difficult.

In asymmetry, we see contrast as a basis for comparing elements in a frame. The principle of contrast stretches to all aspects of visual definition. For example, reading this page would be impossible if the words and the page were both white. A difference between the colors of the text and the paper are necessary to read or, for that matter, to differentiate any element from its environment. Contrast in color, size, position, subject, and shape can be used to build balanced compositions, whether they are symmetrical or asymmetrical arrangements. One must be very diligent in making sure contrast is evident. Elements that are different in size shouldn't be a little larger; they should be much larger than other elements. These naturally derived visual cues provide an avenue for graphic communicators to create visually stimulating designs.

Controlling the viewer's directional flow through a composition gives a designer the ability to emphasize important points. This principle applies to spatial references in detecting closer and more visually distant elements. As we see in nature, elements that create the illusion of depth should overlap and dissolve into an imaginary horizon. The direction of the diminishing shapes lead into the composition, which then connects the viewer to a focal point. Further, contrasting the position of elements in the frame creates differences in direction. By balancing these contrasts, a flow can be achieved that provides the audience with a directional Gestalt. See Color Plate 3-4 in color insert. It is evident that, if the eyes of figures in a composition are looking in one general direction, the audience will similarly look in the same direction. Finally, contrast in the shape of graphic elements can produce a balanced perception of visual weight and direction. Many red, small circles printed across a page can be easily balanced by a large, yellow square. The directional flow toward the yellow square is unmistakably creating a natural focal point for the design. Understanding the perception of visual weight and controlling the directional flow of a composition are avenues in which a designer can travel to create intentional messages.

Figure/Ground

The practice of balancing elements in a visual frame of reference threads through all the principles of Gestalt perceptual psychology. The cloth that holds these balanced threads together is the concept of figures related to a background. Without this contrast, humans would be unable to see anything in a visual sense. In nature, a chameleon mimics the colors of its background for protection, because other animals cannot differentiate between the figure of the chameleon and its background. As simple as this concept may seem, it is a major challenge for most creators of graphic design. Why is figure/ground understanding so complex? The answer can be found in the definition between what is considered figure and what is considered background. The pattern-seeking tendency in human perception is not limited to positive space, or what we refer to as the figure. Viewers have all of the same responses to the negative space, or what we refer to as background, in a composition. Graphic communicators must be aware of the shapes created in the negative space of a composition also. The famous visual riddle of the face/vase (see Fig. 3-3) teaches us that what is initially identified as figure can be switched if an

FIGURE 3-3

Do you see faces or a vase?

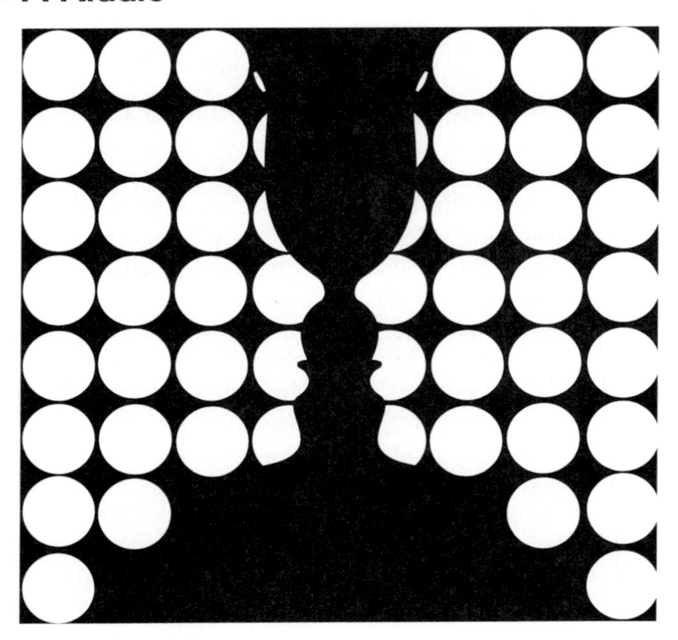

equally recognizable pattern can be detected in the "empty space." Figure/ground relationships can be developed in three main states: distinct, changing, or ambiguous. In a design in which the audience identifies certain elements as figure and the remaining shapes as background, a distinct figure-to-background relationship has been established—words on a page. A changing relationship is demonstrated by our example of the face/vase graphic. The viewer alternates identification of what is figure and what is background between the positive and negative spaces. Our chameleon exemplifies the third type of figure/ground relationship. The viewer cannot be certain of what is considered figure and what is considered background.

Why use any other type of figure/ground relationship other than a distinct one? Today's audience is generally very sophisticated when it comes to visual stimulation. Graphic communicators are challenged to find new ways to entice viewers into giving attention to their designs. Recall when we briefly touched on the ability of special effects producers to suspend reality for a short period of time. As visual communicators, they are attempting to keep an audience interested. In the same vein, using unexpected relationships between a figure and its background generates excitement in an audience. The viewer has to put effort and therefore sustained attention into reading the composition. With all successful visual communication, the intent of the message is critical; however, as we have learned, an intriguing composition may be just as important.

Grouping

Generally, the Gestalt principles based in the concepts of contrast create interest and emphasis in a graphic communication. As a practice, the principles that promote unity temper the principles of contrast. If all the elements in a composition are contrasting, then no emphasis is evident; if everything is different, then nothing is different. Have you ever been bored by too much stimulation? This oxymoron is evident when a designer does not follow the credo of "less is more." Most professional designers agree that, after creating a graphic communication solution, about half of the elements could be omitted and still be effective. A visual communication problem can be compared to solving a children's puzzle for which there is a limited number of pieces and these pieces have specific parameters within a frame. There are no pieces left over, nor can one piece be exchanged for another. The graphic elements in a visual communication should depend on each other to create visual harmony. Each element must have a reason for being included, or it is just an extraneous piecc, not needed in this puzzle.

Keeping this dependency in mind, it is necessary to balance contrast with consistency and emphasis with unity. The Gestalt principles of grouping do just that. The term *grouping* refers to how patterns of similarity are perceived by an audience and remembered by them. Grouping can be achieved through the use of closeness, alignment, repetition, color, shape, direction, subject matter, texture, and anything else that creates the feeling of belonging. One example is family resemblance. You may come from a large family in which each member is very individual in personality, size, shape and coloring, yet some members may be identified as siblings because of one characteristic—perhaps a very similar smile. This similarity can be categorized as a repetition of shape. That one similarity creates a family relationship.

Repetition

One of the weakest areas in Web page design is the lack of unity within pages and within a site. This can be remedied easily by the use of repetitive elements. Repetition is the primary route that can be traveled to attain consistency throughout a graphic communication. When an audience distinguishes a repetitive element in a design, they follow that element as if it were a bouncing ball—witness a bouncing ball prompting viewers to read the subtitles in a movie. Color is the most reliable mode in which to project repetitive unity. Steven Spielberg's movie *Shindler's List* uses this principle in a poignant scene unfolding the horrors of the Jewish ghettos in the late 1930s. The director uses black-and-white footage, with the exception of coloring one figure's coat a bright red. The coat is that of a child whom the Nazis are victimizing. The viewer is compelled to watch from afar as the red coat travels through the gray scenes with the ultimately tragic end showing one red coat piled onto hundreds of black-and-white coats. The repetition of this color and its contrast to all other hues in the scenes focused the audience into watching that one figure. Similar graphic elements repeated in color, shape, size, direction, and texture engenders a consistent following by the viewer.

Proximity

By repeating similar elements, the designer can steer the viewer as well as kindle a sense of unity. The principle of proximity triggers spatial relationships between elements in a frame. The simple fact that the elements are close to each other creates a grouping relationship. If one printed the phrase "less is more" five times evenly spaced on a page, then the audience would identify one group. On the other hand, if

FIGURE 3-4

How many groups do you see in each box?

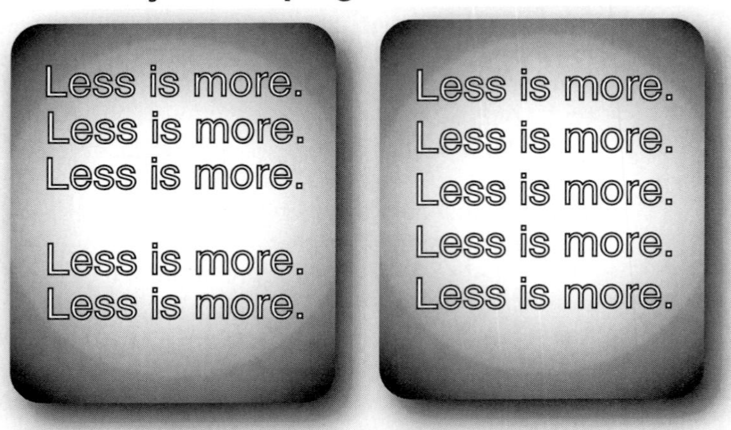

those identical phrases were altered in an arrangement in which two lines were closer together and three lines were evenly spaced, then the viewers would perceive two groups on the page. See Fig. 3-4. Things that are observed as closer together are most often categorized as having a relationship. This is especially valuable when setting text. Text that is composed in forms that are in proximity to each other—closer together—is considered related material.

Alignment

Grouping relationships can be created through repetition and proximity but also can be demonstrated in the principle of alignment. *Alignment* refers to how graphic elements, including text, are justified in a frame. Positioning elements by registering them on an imaginary line within a composition guides a viewer to conclude that these elements are related. Aligning can generate a feeling of ordered categories. It also supports long-term memory cues. Have you ever noticed that the alignment composition of a traditional outline makes it easier to remember large amounts of text? The alternative block text composition gives no hint as to the relationships between the information in the outline. This grouping technique along imaginary lines develops consistency with graphics and text, making a visual communication easier to comprehend and remember.

COLOR THEORY

Our ability to distinguish color is elemental in human existence. Color is so powerful a visual phenomenon that it dominates other elements in the graphic designer's toolbox. Although 85 percent of graphic design is considered related to text, color is the overwhelming thematic construct for that text. Color can be as simple as values of gray on a flyer to the complicated palette of 24-bit depth colors in a computer graphic.

Therefore, it is important for a visual communicator to understand the basics of color theory. Where does one begin to understand such a vast discipline?

Initially, color theory can be broken into two distinct components: physical color properties and psychological color principles. The physical color properties deal with the makeup of color itself. The psychological color principles translate into human responses tagged to certain colors and combinations of colors.

Physical Color Properties

The color primary RGB (red, green, blue) represents the additive set of hues that create the electronic palette for movies, TV, and computer monitors. Electronic presentations also use this primary system. Red, green, and blue light added together in different amounts make up white light—the colors of the light spectrum—hence the name "additive." CMY (cyan, magenta, and yellow) are the subtractive color primaries and are used in the printing industry to create all colors printable. Black (K) is normally added (CMYK) to formulate richer colors in printed pieces. Cyan, yellow, and magenta are subtracted from the additive primaries by using filters, resulting in the primary printing colors—hence the name "subtractive." The third set of primaries—red, yellow, and blue (RYB)—is the set of pigments used by artists and designers to create the spectrum of colors for paintings, illustrations, and paper-based graphics. Red, yellow, and blue are the colors you may remember using as a child to mix a variety of hues for painting and drawing. Depending on the color primary used in a discipline, one can organize the possible color combinations by using a model: a color picker in electronic media, a color wheel in print and fine art media. See Color Plate 3-5 in color insert.

The color wheel allows designers to choose color schemes based on proven successful combinations of hues. See Color Plate 3-6 in color insert. Colors can be identified by three properties: hue, saturation, and value. The hue is simply the color itself. Saturation is the intensity of the hue—the amount of gray in the hue. Colors that are closer to neutral are considered hues with low saturation, and colors that are intensely bright are considered highly saturated. The lightness or darkness of a hue is described as its value—tints, a color with white added to the color, or shades a color with black added. Consequently, the properties of hue, saturation, and value (HSV) can identify colors. The following is a list of basic color schemes available to the beginning graphic communicator:

- *Primary*—any combination of the primary hues
- *Secondary*—any combination of the secondary colors—violet, orange, and green and their tints and shades
- *Tertiary*—any combination of the eight tertiary colors—red-orange, yellow-green, blue-violet, blue-green, yellow-orange, and red-violet
- *Monochromatic*—any single hue and its tints and shades
- *Complementary*—any combination of two colors directly opposite one another on the color wheel
- *Split complementary*—any hue and the two colors next to its complementary color on the color wheel
- *Analogous*—any combination of colors adjacent to each other on the color wheel

When in doubt about which color combinations to use, limiting one's palette is a very effective alternative. It is always useful to remember that applying color should never interfere with the legibility of the message. Color can also be used to delineate a hierarchy of emphasis between elements in a design. Most successful color schemes are limited to a palette of two to four colors with their tints and shades. The practice of limiting colors in a visual communication can foster consistency. Repeating these colors creates unity and a pleasurable aesthetic for the audience. A rule of thumb is to work with percentages of gray to establish value contrasts and then substitute color to add emphasis, clarity, association, and mood. In some instances, it behooves the designer to break this rule and offer a sort of color tension or chaos in a design. Psychological color principles are employed when an audience's emotions are consciously manipulated.

Psychological Color Principles

The properties that make up a color are vital in applying color in a graphic communication, but the color associations created by those combinations are equally important. Advertisers use color to establish a memorable connection between their product or company and the target audience. Even governments use color associations to promote patriotism—the ol' red, white, and blue. It is difficult to separate our perception of the world from how we naturally favor color. Visual communicators have been meticulously aware of the power of color for many centuries. As the visual communication technologies allowed for relatively economical, efficient color application, the study of the psychological effects of color on human physiology expanded. Medical professionals wanted the light blue-green color schemes in their health centers and hospitals to evoke calming emotional responses, while restaurateurs wanted the red colors in their dining areas to stimulate hunger. Some color combinations can even identify eras of history: the neon of the 1960s, preppie blues, the maroons and greens of the 1980s. Regardless of whether these associations are historically accurate or not, a graphic communicator who is cognizant of these prompts can use that information in specifically designed messages.

Generally, reds, oranges, and yellows are connected to warmth, while blues, purples, and greens are associated with coolness. Any color can be made cooler or warmer by its surrounding colors or the complementary color they favor. For example a blue-green is cool, while a yellow-green is warm. Some of the more common color associations can be categorized in the following manner:

- Cool colors tend to recede, while warm colors advance.
- Reds are associated with strength and action. They promote activity, brain stimulation, and excitement. Reds also tend to stimulate aggression, rage, dominance, and rebellion. Conversely, pink (red with white added) triggers the production of epinephrine, used in the body as a tranquilizer.
- Oranges denote bargains, affordability, informality, and cheerfulness. Terracotta (orange with gray added) connotes approachable sophistication.
- Yellow is associated with the sun and is the most readily seen color by the human eye. It is considered cheerful, positive, young, and fresh, but can promote temper and anxiety if used too extensively. Yellow in its gold state can symbolize divinity, harvest, and prosperity.

- Greens are nature's cloaking. They conjure up feelings of restfulness, relaxation, and tranquility. Greens are symbols of nature, growth, and comfort. Dark greens are associated with money in the United States, while dull yellow-green triggers feelings of sickness and infection.
- Blues actually lower blood pressure and heart rate. They symbolize the deepness of the oceans and the expansiveness of the skies. Blue tints and shades dispel feelings of being rushed and harried.
- Purples have historically been associated with royalty. They symbolize pomp, dignity, and exclusivity. They can also be associated with mourning. Purples have religious connotations and can be used to create feelings of properness and tradition.

Both the symbolic and psychological associations related to color need to be addressed when choosing a color theme for a visual communication. In addition, the context of a color's surroundings can alter the way a viewer perceives the value or intensity of that color. This phenomenon is called simultaneous contrast. Michel Eugene Chevreul, a French chemist in the nineteenth century, discovered that colors are altered by the conditions under which they are viewed. See Color Plate 3-7 in color insert. In addition to an awareness of simultaneous contrast, a designer must be confident of the final form in which a color will be output. Red displayed on a computer monitor, a broadcast TV screen, a 35 mm color photograph, an offset press, an inkjet printer, a digital laser printer, and a dye-sublimation printer is unrecognizable as the same color if calibration is not taken into account. Today, graphic communicators are challenged to match color in a variety of viewing conditions and outputs. Respecting the complexities of color design is a giant step in the development of effective graphic messages.

Human perception of color, grouping, and visual organization should determine the technology chosen to convey a message. Historically, the media technology of the time generally drove the development of more sophisticated visuals and applications. The understanding by producers of visual communication that the medium used can sway the meaning of the message intended was paramount in the creation of new technologies. To meet the needs of graphic communicators, new technologies in text and imaging became available. The intent remained relatively constant—to persuade, to inform, or to inspire—yet the means to generate these communications have exploded in variety. Basic design principles, whether used in a simple printed piece or a highly complex animation, can still be employed to produce predictable responses from an audience regardless of the medium. Unfortunately, the mere fact that a message can be sent in so many different modes presents a dilemma of choice for a visual communicator. Which method is the best to convey one's message? How does one determine the media technology needed? In Chapter 4, we will address the enormous variety of visual technologies accessible to the graphic communicator and how to determine which type is best for the project.

BIBLIOGRAPHY AND SUGGESTED READING

Arntson, Amy E. (1998). *Graphic Design Basics*. Orlando, FL: Harcourt Brace College.

Berryman, Gregg. (1990). *Notes on Graphic Design and Visual Communication*. Menlo Park, CA: Crisp.

Birren, Faber. (1970). *Itten, the Elements of Color, a Treatise on the Color System of Johannes Itten Based on His Book the Art of Color*. New York: Van Nostrand Reinhold.
Carter, Rob. (1997). *Working with Computer Type, Color and Type 3*. New York: Watson-Guptill.
Collier, David, and Bob Cotton. (1993). *Basic Desktop Design and Layout*. Cincinnati, OH: Quarto, North Light Books / F & W.
Fisher, Mary Pat, and Paul Zelanski. (1987). *Shaping Space, the Dynamics of Three-Dimensional Design*. New York: CBS / Holt, Rinehart and Winston.
Hall, Edward T. (1973). *The Silent Language*. Garden City, NY: Anchor Press/Doubleday.
Rabb, Margaret Y. (1993). *The Presentation Design Book*. Research Triangle Park, NC: Ventana.
Smith, Stan, and H.F. Ten Holt. (1984). *The Designer's Handbook*. New York: Gallery Books/W. H. Smith.
Wilde, Richard, and Judith Wilde. (1991). *Visual Literacy, a Conceptual Approach to Graphic Problem Solving*. New York: Watson-Guptill.
Williams, Robin. (1994). *The Non-designer's Design Book*. Berkeley, CA: Peachpit Press.
Wong, Wucius. (1987). *Principles of Color Design*. New York: Van Nostrand Reinhold.

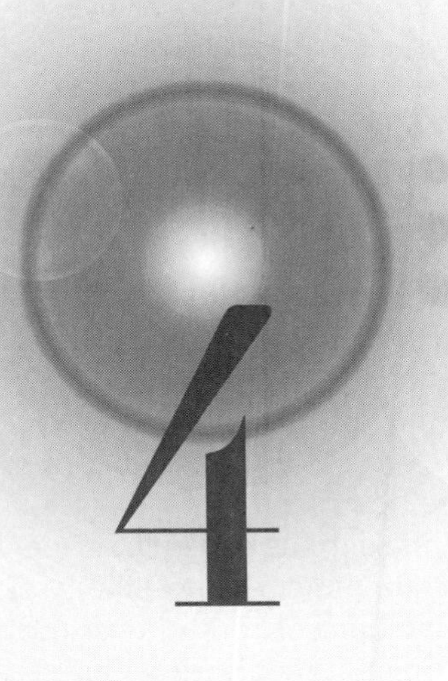

Visual Technology

The Gutenberg shift of visual technology

An engineer, John Presper Eckert, and a physicist, John W. Mauchly, developed the world's first electronic digital computer on February 14, 1946. That Valentine's Day ENIAC (Electronical Numerical Integrator Analyzer and Computer) forever changed the way we communicate. By the 1980s, visual technology in the areas of journalism, marketing, graphic design, advertising, broadcasting, animation, and graphic arts had moved from strictly analog tools to new digital tools. This "Gutenberg shift" transformed not only the way visual communication is created and distributed but also the collaboration of the disciplines themselves. No longer merely a mixture of previously separate disciplines, through convergence these media became a vital new compound. In a compound, the individual particles of different elements chemically combine to produce something unique. In our context, we call that compound multimedia. Multimedia draws the technology map for the future of visual communication by defining new ways of addressing the basic questions "What do we want to say?" and "How do we want to say it?" A thorough discussion of multimedia appears in Chapter 7; nevertheless, the character of multimedia is in convergence of the traditional fields in the visual arts and with the new tools being used in our digital age. Consequently, in this chapter, we will use the term *multimedia* as it applies to print and Web publishing.

Our first step in understanding visual technology is in distinguishing digital multimedia from multiple media. According to Nigel and Jenny Chapman (2000), authors of *Digital Multimedia*, multimedia is "any combination of two or more media, represented in a digital form, sufficiently well integrated to be presented via a single interface, or manipulated by a single computer program" (p. 12). In other words, the integration of media is key. Text, images, and sound can be mixed to produce a multiple sensory experience, in which the viewer can identify and appreciate these modalities as separate entities—for example, as when an audience is viewing a digitally projected presentation while a speaker is gesturing to the screen and narrating. Each aspect of the presentation—text, images, and sound—can be appreciated separately, but these modalities do not actually combine to create a truly integrated experience; hence, this is considered a multiple media presentation.

In a multimedia experience, as in a chemical compound, the audience consumes the media as a cohesive, integrated experience. Mirroring Gestalt perceptual theory, "the whole is greater than the sum of its parts". A strong example of this type of experience can be found on the Internet—specifically, the World Wide Web. A user accesses a Web portal or browser and surfs to any number of integrated media experiences. The user encounters text, images, sound, video, and animation simultaneously and usually has the opportunity to physically manipulate the flow of that experience. This is multimedia, and visual technology is the driving force underlying this convergence. See Fig. 4-1. The opportunity to physically direct the flow of a multimedia session can be shared by both the producer and the consumer, making the experience interactive. Therein lies a unique ingredient in a multimedia compound, the possibility of interactivity—the producer allowing the user a certain amount of control over the media flow. Today, the visual technologies that allow for multimedia experiences are in the hands of producers and graphic communicators learning their craft through new programs at colleges and universities often referred to as media arts, time-based media, multimedia studies, or visual communication technology.

GRAPHIC COMMUNICATION

Graphic communication can be defined as the practice of using Gestalt perceptual theories through visual technology to produce effective designed experiences. In the past, visual technology divided rather than combined media in this area. For example, the graphic arts or commercial printing industry contained about thirty specialties, and errors in the workflow process were constant due to these divisions. Resulting from a

FIGURE 4-1

Notice that, in a multimedia experience, the media stimuli overlap

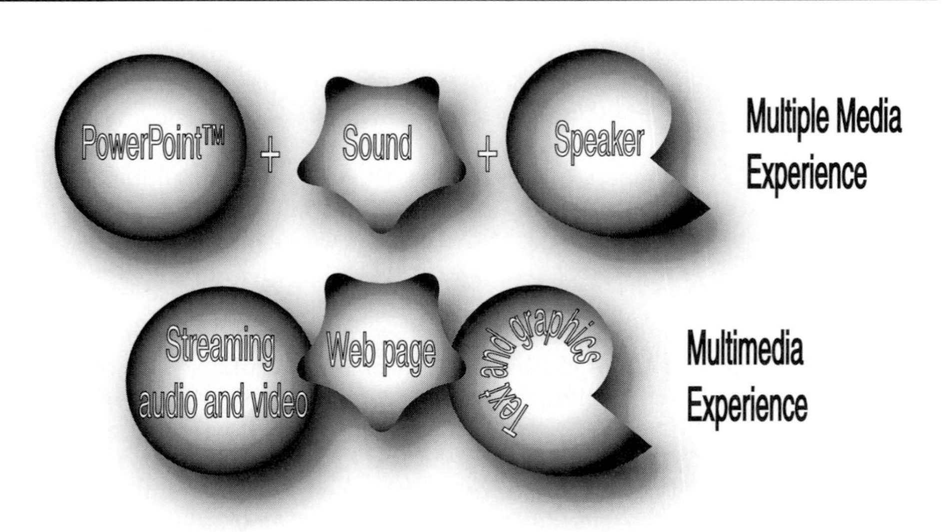

need to streamline the print workflow, automation began with the advent of desktop publishing in 1985, in which ideas, designs, desktop programs, and laser printing combined. The goal was to connect all the digital printing workflow into one vast, looping mechanism, with the client, designer, prepress specialist, and press operator included in the continuum. If we follow this basic map of digital automation combining visual technologies to the present state, the main areas of graphic communication development can be analyzed. This analysis will help us identify the historical pattern of graphic communication disciplines converging to create a new hybrid area, which we'll refer to as media arts.

MEDIA ARTS

It makes sense in this analysis to express visual communication technology advances as a time line of tools. The visual technology used to create graphic communication includes drawing, painting, block printing, the printing press, and finally the electronic computer. The graphic communication discipline has matured into the media arts discipline as the traditional techniques in the printing medium have been replaced by the digital techniques in the multimedia stream. At this point, it may be wise to address our previous delineation that graphic design is limited to two-dimensional elements. It is obvious that, as graphic communication grows into media arts, the latter encompasses not only 2-D spatial elements but also 3-D spatial elements. Three-dimensional space could be, at best, illusionary in a two-dimensional print medium; however, in media arts the paradigm changes. Digital environments allow designers to create actual 3-D spatial elements in three-dimensional space, albeit these graphics are still being distributed in a two-dimensional format—an electronic monitor or screen. Graphic communicators use multimedia visual technologies as a means of integrating text and imagery in still and motion graphic formats to produce a media arts result.

Graphic communication is output through two main distribution systems: the print medium and the broadcast medium. Print includes all collateral distributed using 2-D materials, and broadcast includes all collateral material distributed by electronic means. Print and broadcast distribution can be further divided into analog media and digital media. Analog technology in print, photography, animation, and video are techniques that use continuous signals that produce varying intensities or tones, while digital technology are techniques that use on or off signals or numerical (binary) codes to represent information. A simple representation of the differences between analog and digital information can be demonstrated by comparing a film-based photograph and a digitally based photograph. In a photograph on paper produced by a film camera, the tones are continuous—no dot or pixel pattern is evident. On the other hand, if one looks closely at a digital photograph on a monitor or printout, one sees that the pixels and dots, respectively, are evident in the image. In this chapter, we will limit our discussion to the visual technologies involved in creating print publishing and ultimately Web publishing. The visual technologies involved in creating broadcast media—specifically, television and video—will be addressed in later chapters. Animation or time-based media are unique in that they are a part of all multimedia. Some experts' opinions go so far as to say that all multimedia, including motion pictures, is a type of animation. We will address animation, as it applies to Web publishing, because print information is complemented by the Web medium, and moving imagery becomes an integral part of the translation.

PRINT TECHNOLOGIES

The traditional print industry is going digital. *Cross-media publishing* is the buzzword for the twenty-first century, eluding to the fact that offline and online media are becoming inseparable. *Offline print media* refers to graphic communication that is displayed in pixels, paint, dye, or inks on monitors, paper, board, plastic, fabric, cement, brick, and so on, while *online print media* refers to graphic communication that is displayed in pixels on an electronic screen connected to the Internet. Today, both traditional print output and digital technology create Web output.

To create a context for understanding visual technologies presently being used, we need to comprehend how the traditional print publishing industry has worked since the development of offset lithography techniques. Highly trained artisans through a workflow of designers, prepress technicians, and professional printers controlled the output of printed mass communication. Historically, commercial printing has been in the realm of these graphic artists highly trained in the complicated tools of thermography, flexography, gravure, screen-printing, and offset lithography. A graphic designer developed a text layout with graphics and photographs for the client and then sent the artwork to a prepress technician, who prepared the text, graphics, and photos for the printing press. The prepress artisan translated the designer's work into film, which was transferred onto plates for the CMYK imagesetter. At this juncture, a press operator, an expert in whatever printing method was being used, produced a color proof to be approved by the designer and client. Finally, the corrected, calibrated, and approved design was mass printed, collated, bound, and distributed. This explanation of a typical commercial printing workflow has touched on the main points in this process. Many of the more tedious tasks, such as color matching, are mandatory steps in successful commercial printing. See Fig. 4-2. The point here is to

FIGURE 4-2

A flow chart of the traditional printing process steps

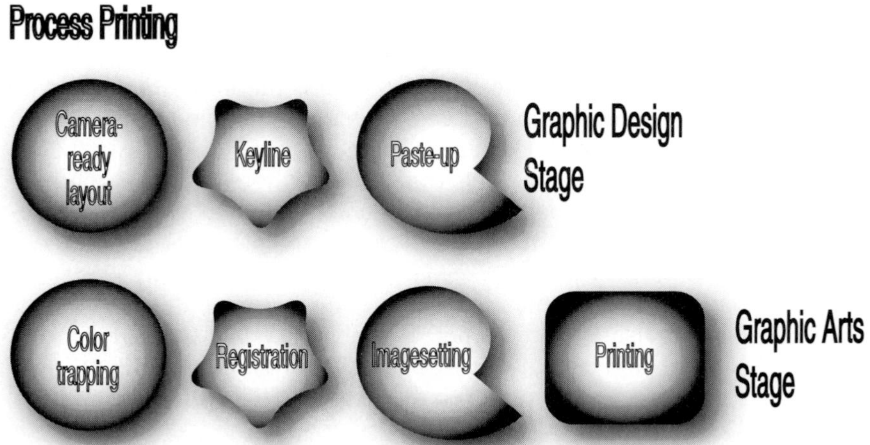

Process Printing

Camera-ready layout
Keyline
Paste-up
Graphic Design Stage

Color trapping
Registration
Imagesetting
Printing
Graphic Arts Stage

emphasize that this process is completed by experts, using very complicated tools, requiring massive amounts of training to master.

DESKTOP PUBLISHING

The traditional graphic arts process involves layout, keyline, paste-up, color trapping, registration, imagesetting, and printing. The entire process was revolutionized in the mid 1980s by the emergence of desktop publishing (DTP). The traditional print processing steps were simplified and significantly compressed by the ability of a software/hardware combination of tools taking ideas from design to camera-ready in a series of simple steps. Even more powerful was the connection of the possible printing outputs. DTP allowed laypeople to publish and distribute flyers, brochures, business packages, cards, and letters in a full-color, professional format. Print communication could actually skip the graphic arts professionals in a short-run job. The quality of output that low-cost inkjet printers brought to the fore outstripped that of the expensive, complicated traditional printing processes.

Desktop publishing brought the obscure tools of the graphic designer, layout artist, keyline specialist, and paste-up professional into one integrated workstation. On the whole, commercial printers still remain analog-based, although, with the improving technologies of computer to plate (CtP) methods, digital online workflow and media-neutral data that can be used in all output options, this is changing. Even further, instead of using separate typesetting systems for editing text and electronic image editing stations for images, professionals today rely on text and image integration through the use of a PC system and software based on Postscript programming language. ICC (International Color Consortium) profiles, which describe standard color behaviors for each device used in the digital print process, can be used for scanners, monitors, and software packages for accurate color calibration in color matching systems (CMS). The professional printer has morphed into a media service provider in a workflow in which a client can initiate data online and send the information to the media service provider, who downloads, manages, and prepares the job and tracks its progress, creating a continuous loop among the client, designer, and media service provider.

This new breed of media service provider can handle a job from the creative design all the way to finishing and shipping. As the industry works on establishing a standard job definition format (JDF) for the entire digital printing workflow using media-neutral data, small and medium-size companies are taking advantage of the less expensive options available within this type of environment. Personalized printing has exploded, as the advances in visual technology have drastically reduced the cost of small print runs that pinpoint specific target audiences. Interestingly, the paper industry reports that consumption has increased during this shift in the industry. Electronic publishing is actually complementing the use of paper communication, rather than usurping it. Because of this move from the general audience to specific target audiences, companies specializing in the minutiae of print collateral are flourishing. For example, Lightning Source, a publishing company located in LaVergne, Tennessee, prints books with a circulation of one, while IDG Books Worldwide allows a consumer to select parts of already published books and compiles the parts into a personal tome. Borders Books retail chain allows its customers to print a book while waiting in the store. It is no longer the size and financial strength of a company but, rather, the company's flexibility and speed that decide market share in the age of printing on demand services.

Similarly, as the printer moves into becoming a media service provider, the graphic communicator is catapulted into the integrated role of a media artist. Graphic designers can no longer isolate themselves from copywriters, marketers, advertisers, or broadcasters. In this age of merging disciplines, viable creative professionals are expanding their training to include graphic communication multimedia in their palette of skills.

BASIC SOFTWARE NEEDS

Along with an expanded skill base comes the responsibility for familiarizing oneself with the appropriate digital tools for the design solving process. As we have previously discussed, electronic techniques are the mainstay of the media artist. These tools can be categorized as hardware, software, or a combination thereof. Hardware is usually defined as the physical presence of such things as the monitor, the central processing unit (CPU), and peripherals, such as the scanner, printer, mouse, keyboard, digital camera, tablet, and stylus. Software is usually identified as the programming code that runs the hardware. Some of these tools defy strict definition in that they are firmware cards or turnkey systems that require both a hardware component and a software component. At this juncture, we will assume that one's hardware, firmware cards, and peripherals are in place. For our purposes, it is inevitable that the hardware used should be platform independent and the choices remaining are purely software-based. There is some debate about a time line for one standard interface for hardware and its structures, but the publishing medium that solves that problem is already in place—namely, the Internet. The very essence of the WWW is platform independence. The arguments about which processing, input, and output devices are best to use become moot in this environment. Therefore, the three pieces of hardware necessary for print and Web design are already standard: a processor, an input device (the mouse, tablet, keyboard, scanner, and so on), and an output device (the monitor, printer, projector, and so on). See Fig. 4-3.

GENERAL MEDIA ARTS SOFTWARE

Media arts software is merely a set of programming instructions packaged to perform dedicated tasks, such as drawing, layout, or word processing. Complications arise when choosing the right program for the job, because the sheer variety of packages can be overwhelming. What does each of these packages do? How can it help solve specific problems? There are virtually hundreds, probably thousands, of software programs available for purchase and use today. We need to sift through them and get to the main categories of programs needed to solve graphic communication design problems. First, it is prudent to realize that buying software can be an expensive prospect, but there are options in this task. Second, software can be purchased in three ways: retail, educational, and as shareware. Retail is obviously the most expensive avenue—software is purchased full-price from a store or catalog outlet or online. The advantage of this type of purchase is that, most often, there is a ten- to thirty-day range of technical support included, as well as all the extensions and add-ons that a particular package offers. Buying software via the educational option requires that the

FIGURE 4-3

The three necessary hardware components for the creation of graphic communication

buyer be a student or an employee of an educational institution. The advantage here is a significantly lower price than retail, and these packages usually include the full software repertoire. Unfortunately, some of the more ubiquitous graphic software manufacturers scale down the educational versions of their programs, excluding some extensions or even the full complement of the program's tools. Educational discounts can be found through an institution's technology division or directly online through the software manufacturer's Web site. Some software comes bundled with the purchase of various input devices, such as scanners and digital cameras. These packages also tend to be the "light" version of the programs. Finally, one can purchase software online in a format known as shareware. Shareware is an independently developed software package that can be downloaded to a user's system for a nominal fee. The advantage of shareware is cost savings, and most shareware is on the edge of creative programming options in its discipline. The disadvantage of shareware is that mainstream software is unavailable, and the developers usually do not support the software—it may become obsolete very quickly. Still, some of the most used programs in the industry started out as shareware. After choosing an avenue of purchase, the next step is to implement a media arts software package to solve a specific graphic communication problem.

Media arts software can be divided into eight main categories: layout, painting, drawing, image editing, animation, Web editing, multimedia authoring, and utility. In each category, a necessary task or several necessary tasks are performed in the digital media arts process. The final output of the solutions drafted from these software programs is displayed through either hardcopy (print media) or softcopy (electronic media).

Some programs have emerged as the clear leaders in their categories, such as Adobe PhotoShop™ in image editing. See Fig. 4-4. Other programs have stiff competition from many software options in their category, as in the case of Web editing software choices, where no package has become the undisputed champion. Nevertheless, it is in the hands of the creative professional to choose the correct tools

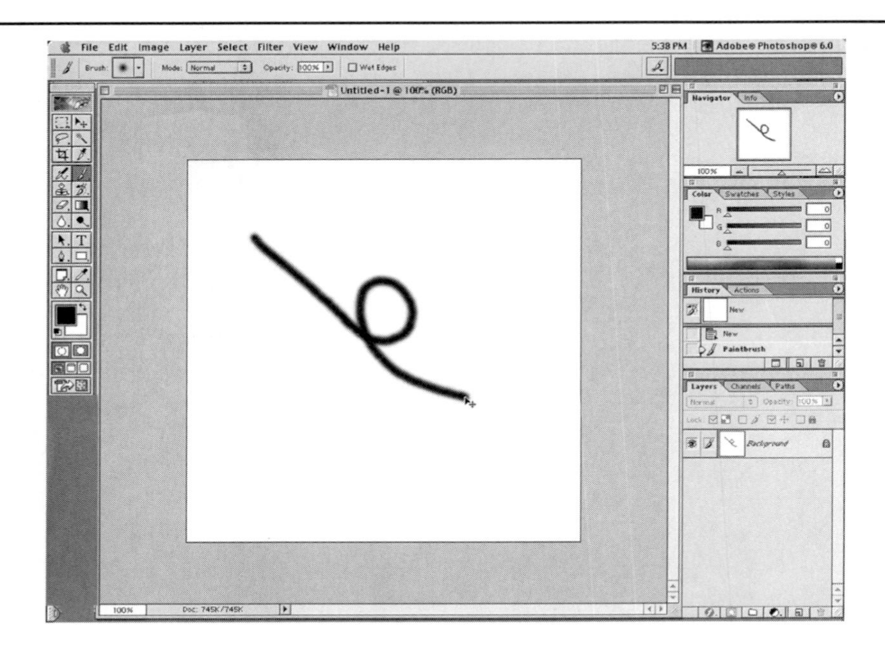

FIGURE 4-4
The Adobe PhotoShop user interface
Source: Adobe product screenshot reprinted with permission from Adobe Systems Incorporated

and awareness of all the available software options is important. In the following paragraph, we will explain each category of software as we would encounter them during the design problem solving process, then we can better understand the value of each type of software. It is somewhat superfluous to describe in text the practice of using graphics software, but at least this avenue provides a starting point for production. A typical design problem would include both print and electronic solutions; therefore, we will follow the process from idea generation to final outputs.

Suppose our job is to create a commercial presence for an imaginary medical database company specializing in the ability to access the medical histories of individuals at any time—day or night. We'll call the company MediData. Our marketing department has gathered the structure and information about the company and has e-mailed all these data to us in the media arts department. We begin by organizing the information into manageable chunks. The copywriters start to develop text for the brochures and booklets that MediData will be sending out to the target audiences: seniors, parents, and travelers. The graphic designers are brainstorming a visual identity that will carry through all the graphic communications the company puts out, employing the principle of repetition to create unity. They are also developing a layout based on a consistent underlying grid structure for MediData's visual communication campaign. Our photographers are taking photos of MediData's personnel and of various identifying areas in the company's headquarters. Photos of the devices needed to implement the program are also being shot. The artists are drawing flow chart illustrations of MediData's medical information retrieval process, artwork for the cover of the print collateral, and images for the Web site homepage. Our image-editing specialists are scanning and manipulating these illustrations and refining graphic metaphors, such as clickable buttons for the print pieces and the Web pages. The prepress artisans are collecting all these components and putting them into the layouts created by the graphic designers. Eventually, the finished layouts will be

FIGURE 4-5

Creative people involved in the MediData campaign

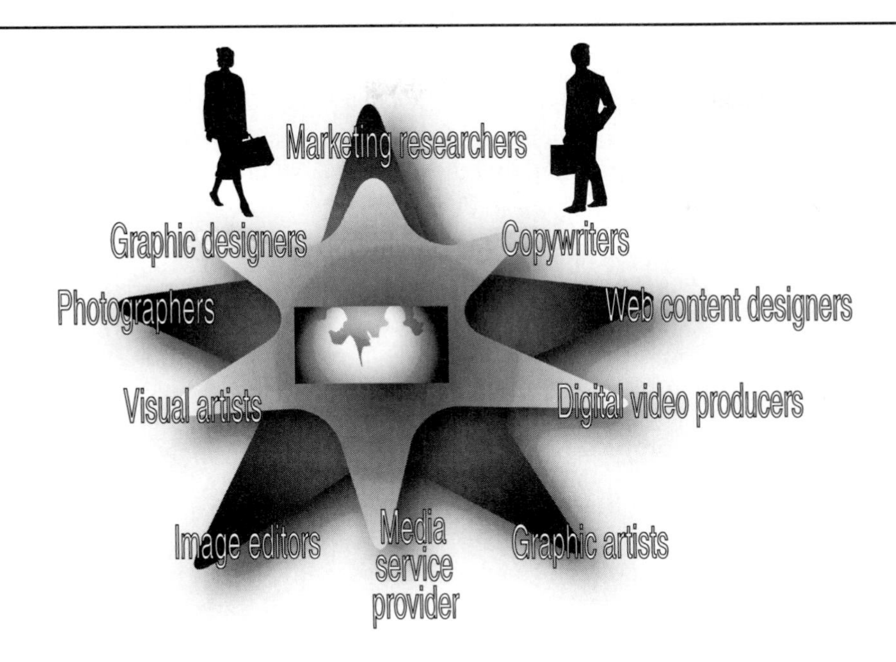

e-mailed to the media service provider, where the artwork will be proofed, imageset, printed, and distributed. The Web artisans are collecting the components pertinent to Internet and CD output. These artists are developing interfaces, including animations, that follow the print collateral grid structures but add interactivity to MediData's electronic presence in the marketplace. The Web specialists will upload the pages onto the Internet and burn the interactive presentations, such as digital video, to the CDs accompanying the printed booklets for mass distribution to the target audience. As one can surmise from this scenario, the steps in the creative process are interconnected. See Fig. 4-5. Cross-media publishing is the result.

Continuing our story, let's look at which software packages need to be employed at each phase in the line of production for the MediData Company. The copywriters use word-processing software that enables the user to input text and establish paragraph formatting in the interface. A word processor can edit text in a nonlinear fashion, unlike the traditional typewriter. Microsoft Word™ is a leading word-processing program, and it has some minor layout capabilities. Text is usually constructed with vectors rather than bitmaps; therefore, the output from a word processor is smooth, scalable letterforms. We will address vector and bitmap distinctions a little later. For now, let's concentrate on the idiosyncrasies of each software category, rather than how it digitizes information.

The graphic designers use a layout program to set up the grid structure that will be used for the entire MediData visual communication campaign. QuarkXpress™, Adobe PageMaker™, and Adobe InDesign™ are three popular layout programs in the graphic arts industry. These packages include full-blown word processors for composing text. In layout programs, designers are able to set up placeholders for text, graphics, and photos by defining areas of the pages as each kind of element. The job of assembling these graphic elements into a cohesive layout can be completed using one of the aforementioned programs. A layout program sets up links to the actual high-resolution digital photographs and illustrations and embeds type within the page itself. Linking large images rather than embedding them in the page itself frees the layout program from redundancy and crashes due to the overextension of memory allotments. Linking can be best explained as a function that points to where the program can find the original large image file. Why save the large image twice—once as the original image and once in the page layout—when linking saves memory requirements with the same quality output? Historically, designers created layouts called blueline boards, everything that was to print in black was pasted-up to a main board. An artwork and text galley, text that was already set with a typesetter and output to photographic paper, was pasted down to the nonphoto-blue board. Anything else was indicated with a letter and number placeholder. The original photos and colored illustrations were marked on the back and sent to the printer in an envelope to be "stripped into" the layout in the film stage of the printing process. This antiquated process has been somewhat maintained in the way layout programs link images rather than embed them into a page layout. A low-resolution proof can be printed on an inkjet or a laser printer to produce highly accurate mock-ups of the final printed pieces. Most often, layout programs are bitmap-based, but the text is vector-based and output to inkjet printers is in a language called PostScript. The Adobe Corporation created a printing language that translates, with great accuracy, letterforms from the binary code of a computer to the dots of a printer. Therefore, Postscript type fonts can be transferred from the computer screen to the laser or inkjet printer in smooth, crisp letterforms, because they are vector descriptions. Adobe also produces a utility called ATM (Adobe Type Manager), which allows non-PostScript printers as well as monitors to display clean, crisp text outputs. It is wise to use PostScript type fonts when the ultimate output is going to be professionally printed, because most media service providers need both the screen font description and the printer font description makeup of a PostScript typeface. It should be understood that digital typography and composition is a discipline unto itself, and this is an extremely simplified explanation of the basic concepts involved in digital text composition.

The photographers and illustrators use image-editing programs to manipulate their existing analog imagery. As mentioned, Adobe PhotoShop is the premiere image-editing package in use today. These types of programs translate analog imagery into digital material via a digital camera, flatbed scanner, or any other analog-to-digital imaging peripheral and allow unlimited manipulation. For example, damaged photos can be restored to nearly perfect condition, or several photos can be merged through layers, resulting in a visually rich collage. PhotoShop is a bitmapped—or, more accurately, a pixelmap—graphics program, because it displays a rectangular array of picture elements (pixels) in a map, each of which is encoded as a single binary digit. Every pixel is mapped at a certain location (address) in small, square dots, which merge optically when viewed at a distance from the screen. When a bitmapped graphic is enlarged or reduced from its original generation, pixels are added or subtracted, respectively. Since the map of pixels is in set locations, colors, and values on the monitor, if the image is enlarged, the processor must interpolate, or make a logical guess, when adding the extra pixels needed to make the image larger. The processor must also interpolate the pixels to clip when reducing an image. The resulting image usually shows aliasing, or ragged edges because the pixels start to gather into what looks like rough blocks. A function called anti-aliasing battles this effect by creating a blurred edge, so that the pixels seem smoother.

At this point, let us address the concept of resolution. Each hardware device in this process has its own resolution: most monitors display 72 ppi (pixels per inch) to 96 ppi; scanners and digital cameras display at 300 dpi (dots per inch) to 3,600 dpi; average inkjet printers can define 300 dpi to 800 dpi; offset lithographic printers can create more than 2,400 dpi at 400 lpi (lines per inch). As one can see, the measuring parameters change with different outputs. To simplify, digital images should be defined exclusively in pixels per inch. A 288 x 360 ppi image on a 72 ppi monitor is 4 x 5 inches in actual size. If the same image is output to a 300 dpi inkjet printer, its output size is still 4 x 5 inches, but its resolution is finer, with 300 pixels packed into each inch, rather than just the 72 pixels showing on the monitor. In addition, some confusion arises when saving an image at certain resolution levels.

Generally, the higher the defined resolution of a digital image, the larger the image and the more space it occupies in storage. A good rule of thumb is to match image resolution to the output device. For example, if we wanted to scan an analog photograph and show it in on a Web page, it is only necessary to scan at 72 ppi to 96 ppi, because this setting is the maximum pixel array for most monitors (the output device). Further, if we wanted to use the same scanned photo but enlarged to twice the original size and to send it to our media service provider to include in a magazine layout, then the parameters change. We would need to calculate the final output of our photo and base our resolution on that number. Screen output resolution will be adequate if set at 72 ppi, newspaper output resolution would be acceptable at 170 ppi, and magazine output resolution would be smooth at 300 ppi to 400 ppi. Consequently, our photo should be scanned at 300 ppi. Unfortunately, we are also enlarging 100 percent from the original photo, so we need to take this into account when resolution is calculated. The basic scan resolutions are calculated by a 1:1 ratio between original artwork size and final output size. Since our scan needs to be enlarged, it is necessary to add 100 percent more resolution. Therefore, our final scan resolution for the enlarged photo should be set at 600 ppi. If we wanted to reduce 100 percent, then the scan resolution should be cut in half, or 150 ppi. Many formulas have been offered to make these calculations simpler, but in the end matching the resolution of the digital image to the final output resolution is often the most effective, most straightforward method. Keep in mind that storage file size is directly related to resolution and that higher resolutions require enormous amounts of storage space, especially for graphics and multimedia elements.

Returning to our explanation of image-editing software capabilities, the layers in PhotoShop allow for many images to be viewed on top of each other; each layer is independent of the other but seen as a whole. See Fig. 4-6. This is valuable, because, if bitmapped elements were drawn on top of each other in the same layer, then the pixels underneath would be obliterated; in a separate layer, the pixels remain intact. It may be helpful to think of bitmapped graphics as actual paint and canvas. If an artist applies paint on top of existing paint on the canvas, then a mixture of colors result; however, if the painter uses acetate layers and applies paint on each acetate layer separately, the elements remain intact and are viewable as a whole. In the digital world, layers can be either opaque or translucent, providing for the virtual mixing of colors yet maintaining the individual integrity of each layer.

The artists and illustrators use painting and drawing programs to produce original artwork. Digital paintings that emulate natural media, such as oil paint and watercolor, can be produced in a program called Painter™. Painter has had at least three owners since its inception in the early 1990s, nevertheless, it is unsurpassed in its natural

FIGURE 4-6
Layers in a bitmapped graphic

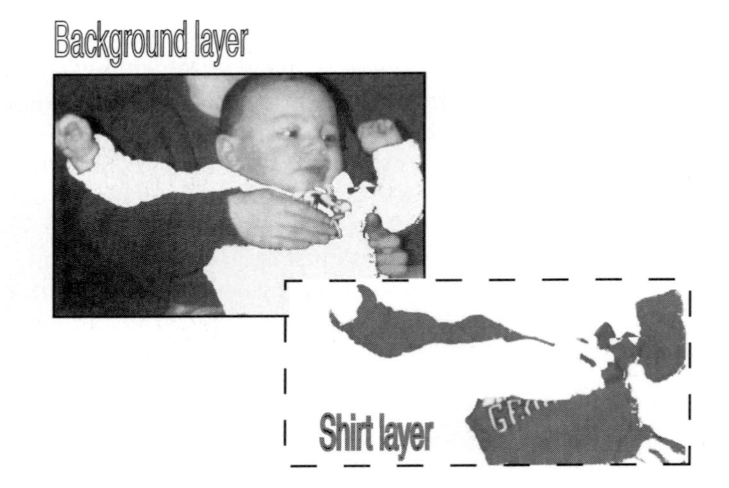

media metaphors. Kai's Power Tools™ and others have small followings, but Painter remains the leader in digital paint programs. These software packages are bitmapped-based but use PhotoShop-like layers to enhance their capabilities. The difference between a paint program and an image-editing program is their inherent emphasis. Image editors concentrate on the manipulation and restoration of existing imagery, while paint programs create imagery from scratch. There are image-editing components in paint programs, but they are limited in comparison with the scope of digital painting tools. It would be fairly difficult to detect whether a painting had been created with real paint or virtual paint when it is output as a commercial printed piece.

Meanwhile, our technical illustrators at MediData use a vector-based—or, more accurately, object-oriented—drawing program to produce hard-edged imagery. We mentioned that the illustrators might be working on drafting flow charts depicting the way medical information is retrieved by MediData. Hard-edged artwork and headline types are best developed in drawing programs such as Adobe Illustrator™ and Macromedia Freehand™. Vector artwork is stored as mathematically defined descriptions of individual paths rather than an array of mapped pixels. Vectors are necessarily displayed as pixels on the monitor screen, but the array is interpreted by computations of the formulas for the lines, curves, and shapes making up the final image on the screen. It is scalable and forgiving in that enlarging and moving a vector does not degrade the image or disturb any underlying elements. See Fig. 4-7. Layers are not required, because each vector element is unique and separate from other elements on the screen. For example, if vector type were laid on top of a bitmapped landscape photograph, then the type could be moved indefinitely without altering the photo beneath. Vectors are like paper cutouts, in comparison with bitmapped graphics, which are like wet paint. Still, most vector programs have layering capabilities. Inevitably, bitmapped graphics require more RAM (random access memory) and storage space than vector graphics, which tend to be compact and less RAM-intensive. Very often, vector drawings and bitmapped paintings are combined in image-editing programs to produce unexpected, illustrative results. Unfortunately, as competition among these dedicated software graphic programs increases, their manufacturers believe that incorporating vector capabilities into bitmap

FIGURE 4-7

Notice that the bitmap image "pixelates" when it is enlarged, while the vector image, although less detailed, is infinitely scalable

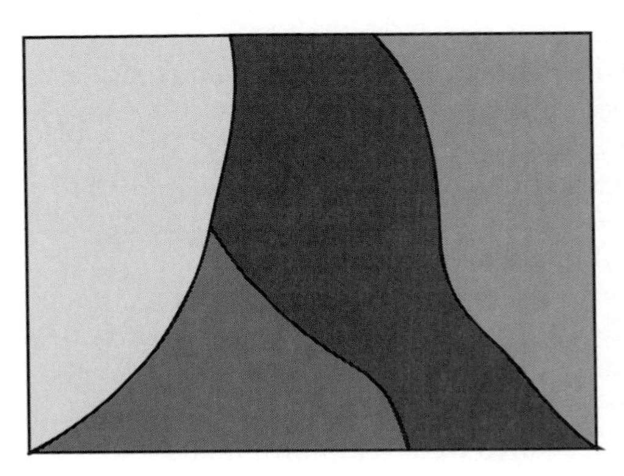

programs and vice versa makes for a more flexible environment; in reality, it tends to weaken the primary strengths of each. On the other hand, adding Web graphic components enhances the usefulness of vector draw programs and bitmapped paint programs.

Our Web producers use a combination of image-editing, painting, drawing, and Web editors to complete the Web site for MediData. The Web specialists manipulate the print information to conform to the unique needs of an online communication environment. Hyperlinks allow for nonlinear navigation in a Web site, and the designers developing MediData's Web presence address this added component. Web designers use HTML to structure text for uploading to the WWW. This convention of tags allows developers to structure text and later imagery in a limited set of layout configurations. An international group of volunteers, the World Wide Web Consortium (W3C), oversee the upgrades and development of technologies for the WWW. Presently, new conventions—including layers and streaming audio and video, to name a few—have been added to the possible layout options open to Web artisans. The WWW is changing rapidly from a strictly text-based medium to a multimedia mecca. Web-editing software has progressed from simple HTML mark-up programs to drag and drop, full-blown assembly packages. What desktop publishing did for the print publishing industry, Web editors are doing for the Web-publishing arena. Web designers are able to employ Macromedia DreamWeaver™, Microsoft FrontPage™, or Adobe GoLive™ to lay out pages that include not only text and graphics but also animation and video, while not having to manually write one line of HTML code. Just as print layout programs such as QuarkXpress generate all the PostScript language code needed for the printer to create a publication, programs such as DreamWeaver generate all the underlying HTML coding needed to view a page on the WWW. This is not to say that a basic understanding of PostScript or HTML is unnecessary but, rather, that a graphic communicator need not have a high level of programming skill to produce sophisticated print and Web publications.

In close collaboration with the Web design communicators, our authoring and editing experts produce our interactive CD, including a digital video. Programs such as Macromedia Director™ and Apple Final Cut Pro™ provide interfaces for the authoring and editing. Director is an authoring package with which artisans can develop interactive games or training sessions and can burn these mini applications to a CD for distribution. Let's say that MediData requires its customers to be adept at using electronic metaphors for accessing their medical files. Our authoring designers are creating an interactive training session on CD, teaching clients how to use these metaphors. A digital video of MediData's CEO explaining the importance of this service is included on both the Web site and the CD. The Web site has a pared down version of the video, while the CD has a full version available for viewing. Low-cost software packages, such as iMovie™, offer amateur videographers the opportunity to produce digital videos with special effects previously exclusive to professional-level digital editing programs. Apple Final Cut Pro and Adobe Premiere™ are professional-level digital video editors and include many levels of video manipulation capabilities not found in the lower-end packages. Nevertheless, desktop video is forging a future market of lay producers and consumers, just as desktop publishing and Web editors have for their respective media disciplines.

MediData has decided to include an animated logo at the beginning of its digital video for the Web site and CD. The animators use vector- and bitmapped-based software programs to develop this animation. In the logo animation, the multimedia designers employ Macromedia Flash™ software for both the digital video and the Web site. Since Flash is vector-based, the subsequent files are compact and work well with the size constraints inherent in the Web environment. Flash uses vectors and scripting to create moving points, lines, and shapes, which are then translated from the native flash file format (.fla) to a shockwave flash format (.swf), which is readable on the Web. At this juncture in the reality of the WWW, viewers of this type of animation still need the proper translator or plugin for their browser to display the flash animation. This scenario is rapidly changing as browser developers include standard animation translators such as shockwave flash directly into their programs, rather than as a third-party plugin.

The designers intend to use Adobe ImageReady™, a bitmapped animation program, to create animated buttons for their electronic media needs. ImageReady also generates compact files. Although it is not vector-based, this program uses very small bitmapped graphics in a frame-by-frame animation format in which the images are highly optimized. The program accomplishes this compressing task by various means. ImageReady uses indexed color, rather than the full palette of colors available. Indexed color saves the hues actually used in the image and discards any other colors in the basic 216-web color palette. In addition, any graphic element that is used more than once in the animation is saved to a kind of library, where it is linked to the actual animation. This technique, also used in Flash, creates the graphic element once and then repeats it, thereby reducing enormous amounts of processing power and file size storage. (The specific mechanics of digital animation will be further discussed in Chapter 8.) Once the animation is produced, the Web designers and digital video editors can include it in their productions. Finally, any problems that arise in the

FIGURE 4-8

The web of software in the MediData campaign

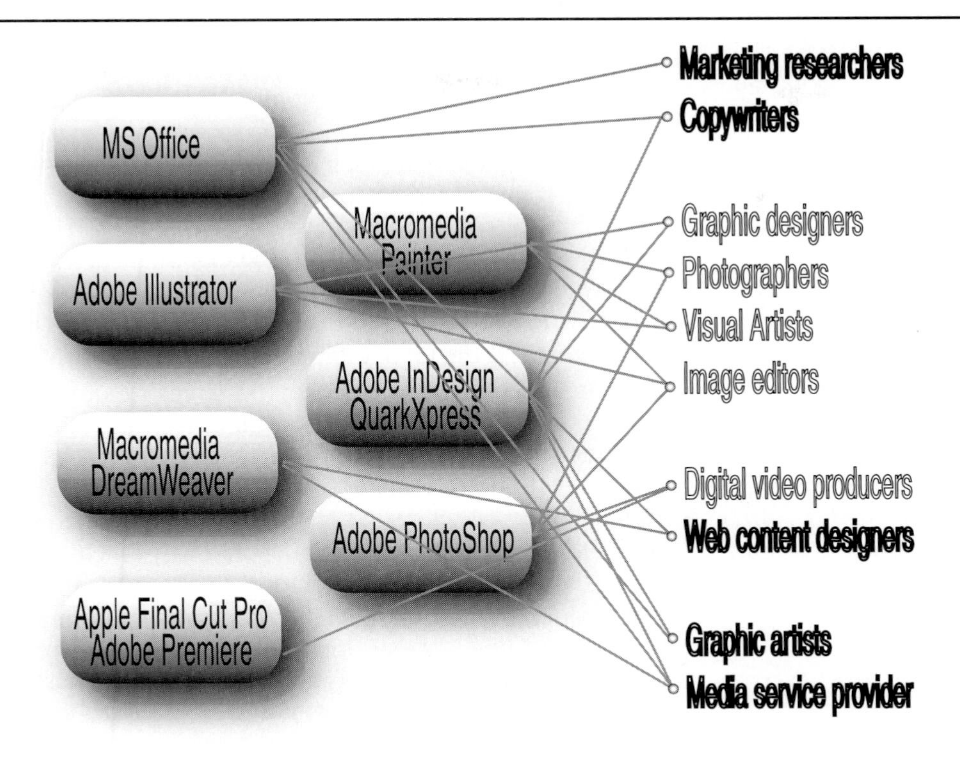

organization of fonts, extension problems, file formatting, and general protection of the software and processor can be solved by utility software. Some utility programs include Norton's AntiVirus™, Cassidy and Greene's Conflict Catcher™, Semantics' Font Suitcase™, and Retrospect's Backup™. See Fig. 4-8.

GRAPHIC FILE FORMATS

Each type of software package we have discussed creates files in its own native format. Depending on final output, the file, whether it be a still graphic, digital video, or digital animation, must be translated into a readable composition for that particular media. All digital files start out in formats unique to the program that created them. For example, PhotoShop begets .psd files; Flash generates .fla files. The following is a list of universal file formats not specific to any particular application but available to most of the graphics software programs:

- TIFF or .tif (Tagged Image File Format). This is a bitmapped file format that is fairly universal for creating lossless quality color graphics. TIFF format is the most stable image file for saving full-resolution graphics. It can display millions of colors and is usually used in print graphic programs, such as PhotoShop and Painter. TIFFs tend to be very large files and require significant amounts of storage space.

- EPS or .eps (Encapsulated PostScript). This is a vector file format that contains PostScript code for the printer and sometimes an optional PICT or TIFF image for screen display. EPS is commonly used for moving files from one application to another. It is also a popular format for color separation. It is often used for clip art.
- JPEG or .jpg (Joint Photographic Experts Group). This is a bitmapped file format but also incorporates compression. Because of this compression capability, JPEGs lose bits of information each time they are closed or opened. This is called lossy compression, as opposed to lossless compression, as in a TIFF file. JPEGs are used for photographic images on the Web due to their small file sizes.
- MPEG or .mpg (Motion Photographic Experts Group). This is a sibling to JPEG, except that it is the format for motion graphics and video.
- PICT or .bmp (Picture or Bitmap). These files are bitmapped file formats. PICT is Macintosh-based and .bmp is Windows-based. Both formats are used to display and store images. They are the preferred format when opening a graphics file in a motion or an animation program.
- DCS or .dcs (Desktop Color Separation). This is a file format for saving a CMYK image for color separation, with the options for including spot color channels, alpha channels, and a low-resolution file for previewing.
- PDF or .pdf (Portable Document File). This is a format that can link and/or embed bitmapped images, fonts, and vector images into one file. It has facilities to include hypertext links and audio/video and has compatibility with the WWW. The PDF format was originally developed by Adobe Systems to provide a portable file standard that would work in any program or platform. A program called Adobe Acrobat Reader™ exclusively displays PDFs. Acrobat Distiller™ displays PDF files in a PostScript format and Adobe Exchange™ allows producers to include audio/video. Adobe Acrobat Reader is distributed for free, but all the other components must be purchased.
- GIF or .gif (Graphics Interchange Format). This is a bitmapped format used mainly for hard-edged graphics on the Web, such as clip art. It can support up to 256 colors. GIFs compress to low file sizes and are accepted by all browsers. Although, .gif is bitmapped-based, most artwork produced in vector draw programs are translated to GIFs for displaying in Web pages. GIF89a is a GIF file that supports transparency and flash animation. Software developers are required get permission from the CompuServe company to use GIF formats, because this organization holds the patent.
- RTF or .rtf (Rich Text Format). This is a vector format used to transfer text from one application to another. It retains the original style of font faces and keeps the text formatting intact.
- PNG or .png (Picture Network Graphic). This is a format created to replace the CompuServe GIF and JPEG. It is patent free and has lossless compression capability. Generation 4 Web browsers support PNG.

There are many more formats, but this list is a compilation of the basic files that have emerged as standards among creative professionals. Transferring files between programs has become common practice, requiring portable files to retain their integrity in text and imagery.

THE ART OF VISUAL COMMUNICATION

Our extended example of the imaginary company MediData took us from idea generation through graphic construction to publishing a media campaign. The basic software requirements were discussed as individual entities, but optimally the experts in collaboration at each step in this process would use these separate tools. An even greater shift in pedagogy occurs when one or two media arts professionals replace individual artisans and software experts. The invention of visual technologies that permitted laypeople to create highly sophisticated layouts in various venues changed the way mass information can be published and consumed. It is possible to use these software packages and thereby produce mass communication without extensive training or theoretical understanding. The democratizing of graphic production has both enhanced and degraded its results. Enhancements come in the form of general availability, affordable personalization, and flexible media. The degradations in the process come in the practice of design by uninformed, untrained, and careless producers. This deterioration will be addressed more thoroughly in Chapter 9. One can only hope that, with ease of use, responsibility in production closely follows.

The true art in visual communication is what one leaves out. In other words, it is possible to involve every type of sensory stimulation available by using these new visual technologies, but is it the most effective way to communicate an intended message? As in any simple figure/ground problem, the solution is in balancing the perception of positive elements with the negative elements. It must be remembered that the shapes perceived in the background areas equally impact the audience. Similarly, bombarding one's audience with information translated in every media venue available may not be the best solution to a communication problem. The emphasis is lost. Effective design in visual communication is like a musical moment. Memorable music is created in the pauses between the notes. Powerful graphic communication is ultimately a symphony of choices between what to put in and what to leave out.

BIBLIOGRAPHY AND SUGGESTED READING

Abdulezer, Susan. (2001). *The State of the Arts: Curating the Digital Classroom. Converge: Education, Technology*. Folsom, CA: e.Republic.

Baldwin, Carliss Y., and Kim B. Clark. (2000). *The Power of Modularity*. Cambridge, MA: MIT Press.

Bistrich, Andrea. (2001). Overcoming Barriers. *Print Process: Change in Our Industry*, 13(1), [online: www.printprocess.net/index.asp].

Chapman, Nigel, and Jenny Chapman. (2000). *Digital Multimedia*. New York: John Wiley & Sons.

Cottrell, Leonard. (Ed.). (1998). *Print Publishing Guide: The Essential Resource for Print Publishing*. Indianapolis, IN: Macmillan Computer Publishing.

Fleischman, John. (2001). *Creative Tools for Creative Programs. Converge: Education, Technology*. Folsom, CA: e.Republic.

Hendricks, Bernd. (2001). One Click to Print. *Print Process: Change in Our Industry*, 13(1), [online: www.printprocess.net/index.asp].

Petz, Susanne. (2001). Entangled Paths. *Print Process: Change in Our Industry*, 13(1), [online: www.printprocess.net/index.asp].

Romano, Frank. (2001). Systems Thinking. *Print Process: Change in Our Industry*, 13(1), [online: www.printprocess.net/index.asp].

Tuckett, Simon. (2001). *From Sky to Sea*: A 3D *Journey*. *Step by Step*: 2001 *Illustration Annual*. Peoria, IL: Dynamic Graphics.

Waldman, Harry. (2000). *Computer Color Graphics: Understanding Today's Visual Communication*. Sewickley, PA: GATFPress.

Weaver, Marcia. (2000). *Visual Literacy: How to Read and Use Information in Graphic Form*. New York: LearningExpress LLC.

5 Audio Technology

Do you hear what I hear?

Audio technology is a traditional technology that continues to grow and reinvent itself on many levels. It began with Guglielmo Marconi's telegraph and has evolved into digital audio used on the Internet, in films, and in broadcasting. Audio is often a forgotten element when designing media and misused by many. The key to utilizing audio technology properly is in comprehending the process of producing audio from beginning to end. This is not an easy task, since audio technology is being updated on a daily basis. See Fig. 5-1. However, the basic steps have remained somewhat intact, even though the equipment being used may have changed slightly. The entire process begins with learning what sounds to record and how to record them and, if they are already recorded, how to transfer the audio to your production. The next steps in the process are to mix and edit the audio and, finally, to place the audio into a setting for playback. There are several ways this can be accomplished, from creating streaming audio for a Web site to developing audio for a television commercial. The end uses of audio technology can vary a great deal and usually depend on which technology is dominating society at the time. The Internet, radio, and television are the dominant technologies of the day and will be the focus of sound communication design.

Sound Communication

Sound communication design begins with the selection of the type of sound needed for the production. Does the production require narration, music, sound effects, or sound on tape (audio with video)? The selection of the audio can make a Web site come to life or irritate the end user by endlessly looping for no reason. Think about the many times you have been surfing the Internet and the music from a site plays over and over until you stop it or just leave the site out of exasperation. This is poor sound design. Sound should enhance and even help the end user understand the message better. Also, it is important what type of sound you select for your media

FIGURE 5-1
Audio equipment

production. If you are producing a Web site, the types of sounds that you might choose include sound effects for links, music for background, sound effects, or narration for a particular page. If you are producing a video production to be used in a presentation or on television, you might choose background music, narration, or synchronous sound with the video. Either way, you select sound that compliments the message in such a way that the message increases in value to the listener. Once you have chosen the type of sound to be used in the production, you need to record it.

AUDIO PRODUCTION

The media producer has two choices when recording sound. The first choice is to record original work or to use prerecorded sound and pay for the rights to use it. Let's discuss the process of recording original sound and then look into the prerecorded material.

A media producer can record narration, music, and other types of sound on tape through the use of microphones. The quality of the microphone has a great deal to do with the quality of the sound recorded. A current problem with audio produced on the Internet is that some of the sound has been recorded using the computer's internal microphone, which is not of high quality. If you are going to record original sound, it is best to use high-quality equipment and to understand how the equipment works. Microphones should be chosen for use according to their various properties: how they record sound, the type of sound pattern used (such as cardioid, or heart-shaped), and the physical use of them. The two types of microphones most used in production are the condenser and dynamic. The condenser microphone converts sound waves into an electronic/digital signal by using electronically charged plates, which move when someone speaks. This microphone is more sensitive to high-frequency sounds and considered a superior piece of equipment. The dynamic microphone is more rugged and uses a diaphragm attached to a coil, which vibrates when someone speaks, converting the sound into electric current. This microphone can record high-quality sound in difficult conditions, such as at sporting events.

FIGURE 5-2
Microphone patterns

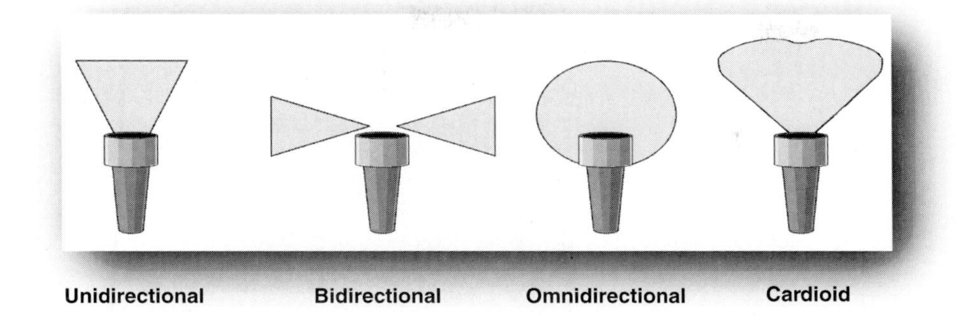

The second way to decide which microphone to select is by the sound pattern. Both the condenser and dynamic microphones have a variety of sound pattern options. The microphone can be unidirectional, which picks up a cone pattern in one direction, or bidirectional, which picks up sound in two directions. Omnidirectional microphones pick up sound from all directions equally well. Cardioid microphones have a pattern that is heart-shaped and can equally pick-up sounds from the sides better than from the top of the microphone. See Fig. 5-2. Supercardioid microphones have a highly sensitive narrow pattern with a long range and are very directional. These microphones can make sounds far away seem very close and are used at football games to increase the feel of being there. Until recently, each microphone had only one pattern, but new microphones now offer several pattern options. You can use a condenser microphone that has unidirectional, omnidirectional, and cardioid patterns, which can be selected with the flip of a switch. It is best to use the pattern that suits the recording setting. For example, use a dynamic unidirectional microphone when interviewing a person outside near a road. This selection will ensure high-quality sound with little background noise.

The third distinguishing characteristic of microphones is how they are physically used in a situation. There are stationary and mobile microphones. See Fig. 5-3. Obviously, the stationary does not move around much, while the mobile microphones are moved a lot. Mobile microphones include lavaliere, hand, wireless, and boom microphones. Lavaliere microphones are the small tie clip microphones used in news. Hand, or stick, microphones are the hand-held devices used for interviewing or for vocalists at concerts. Wireless microphones have recently become popular and free the user from cables. This microphone sends a signal to a receiver, which transfers the signal to a speaker or recording device. Wireless microphones can be lavaliere, hand, or headset, depending on the user's desired look. The boom microphone is used by hanging it over the action being recorded; it gives the talent total freedom from wearing any wires at all. Stationary microphones include desk, stand, and hanging microphones. Desk microphones are used in interviews and panel discussions and do not usually move but, rather, are placed in front of the talent on a desk or table. Stand microphones are employed when the user cannot move around a great deal; they add to the look of the performance, such as at a rock concert. Hanging microphones are used when a boom would be impractical and the talent needs to be free of wearing wires. It is important to remember that this is just one way of

FIGURE 5-3
A stationary microphone

classifying microphones, mobile microphones can be used as stationary and not all stationary microphones are fixed to one location.

Recording high-quality audio is the first part of the process in audio production; the next is the mixing of the various sounds selected for recording. Audio mixing can take place when the sound is being recorded or after it has been recorded. Media producers can have the microphones patched into an audio console that takes several inputs (CD player, mini disc, MP3 player, and so on) and mixes the various signals into one or two audio tracks. The console controls the volume of each source, allows the mixing of the sound, lets you shape the character of the sound through equalizers/special effects devices, monitors the quality of the sound through the VU (volume unit) meter, and sends the sound to output sources (speakers and recording devices). The audio console is used in studio recording; however, if a producer is out in the field recording audio, a less complex mixer can be utilized. The media producer can use a passive mixer that allows several microphones to be patched into the recording device or can use an active mixer that also allows volume control. This is mixing the sound as it is recorded. Sound can also be mixed and edited after it is recorded. Sound can be recorded in analog or digital signal and on many formats. Analog signals are sound recorded as electrical impulses on a tape. Digital signals are sound recorded as a pattern of 1s and 0s; digital has higher quality than analog and can be recorded on a cassette tape, CD, mini CD or computer. See Fig. 5-4. The analog signal can also be converted to digital for editing purposes.

The devices used for recording sound range from the old analog tape formats to digital formats on a computer or portable digital recorder. Portable analog cassette recorders are still being used in audio production, and the electronic signal is then converted to a digital signal for editing. The entire audio production industry has converted to using digital equipment, such as the digital audiotape recorder (DAT), computer disks, mini disc (MD), compact disc (CD), and digital versatile disc (DVD). The DAT records high-quality sound with time code, high-speed search, and cueing abilities, as well as portability. Sound can also be saved to a computer's hard disk, zip

FIGURE 5-4

A cassette, a CD, and a mini CD

disk, and jazz disk. This is assuming that the computer has the necessary software and hardware to record audio, such as Pro Tools™ software and a sound card. The sound quality can be changed to any level desired by the media producer, and computer hard drives can save large quantities of audio, depending on the compression software. One minute of uncompressed CD quality sound with stereo capabilities takes up about 10.5 megabytes of space on the hard drive. This can fill up a hard drive quickly and is the reason that portable disks have become so necessary. The zip and jazz disks can also save audio files and make it easier to exchange the files with other computers and save space on hard drives for other software plugins.

The mini disc is small, about 2 ½ inches, and can store one hour of high-quality stereo sound. The compact disc was used for playback only, until recently, when read/write disks were invented. These discs can record and play back high-quality audio and store about 650 megabytes of data, or eighty minutes. The digital versatile disc, also known as digital video disc, is similar to the CD but can store 4.7 gigabytes of high-quality video and audio, or more than two hours of broadcast-level data. The data are stored in a pattern of 1s and 0s, which are converted to a series of pits on the disc and then read by a laser stylus to play back the sound. This is the basic configuration of the CD/DVD system. The DVD has been playback only but is quickly moving toward read/write format.

Once the sound is stored, the media producer can integrate the audio directly into the end product. For example, some narration recorded and stored on a hard drive can be converted to an audio format file known as QuickTime or WAV or AIFF files to be used in a Web site. However, the media producer would most likely take the recorded audio to a digital audio workstation (DAW), a computer with audio-editing software, such as Sound Forge or Pro Tools, which mixes together multiple tracks of audio into audio files to be used with video or alone. See Fig. 5-5. These audio workstations not only allow one to edit and mix the sound but also to save the final audio clip into the numerous file formats being used in today's "no standard" society.

FIGURE 5-5

A digital audio workstation

The typical equipment found in a DAW consists of a computer with CPU, a monitor, a keyboard, a mouse, an audio console, an amplifier, a mini disc player, a compact disc player, speakers, a DAT, and microphones. These machines give the media producer several options for mixing together various sources and outputting them to your format of choice. The DAW may also have a DVD or other equipment that might be necessary to the production, but the basic hardware setup usually has the equipment listed. The computer can be a PC or Macintosh but needs to meet the minimum requirements of the software you choose.

There are literally hundreds of audio-editing software programs available to the public today, including the simple sound recorder software packaged with every computer sold since 1996. The two most used software programs in DAW systems are (PC) Sound Forge™ and (Mac) Pro Tools™. PC-based DAWs also use SADiE™, Soundscape™, WaveFrame™, programs while Mac based systems use AudioVision™, and Sonic Solutions™ programs. Sonic Foundry's Sound Forge and Digidesign's Pro Tools are the dominant digital audio-editing software, because they have a great deal of versatility. See Fig. 5-6. These programs can mix as well as edit audio tracks and then save the audio into several different file formats, which can make the life of a media producer much easier. Let's discuss what takes place in the actual editing of sound in a DAW.

The editing software opens up an audio session with multiple windows with several functions, such as an edit window, a transport window, and a mix window. The edit window is where you build different audio tracks, which can be as many as fifty-five tracks or as few as one. You can record new audio in each track by delegating where the audio is coming from on the audio console. The audio can be music or voice fed through the audio console to the computer or can be recorded directly onto the computer. In the edit window, you can record sound, play sound, trim the sound by note or by word, amplify the sound, repeat the sound, mute the sound, fade the sound in and out, and add different layers of sound. This window also allows you to drop and drag audio clips into any order. The transport window is a set of icon buttons that let you play, rewind, fast-forward, search, and record sound. The mix window brings up

FIGURE 5-6

A Pro Tools screen capture
Source: Screenshot reprinted by permission from Digidesigns, Inc

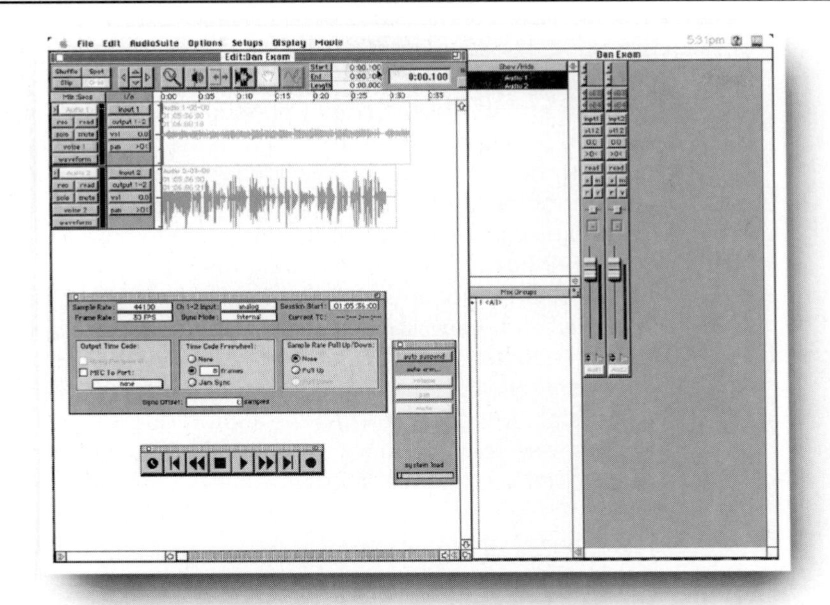

a digital audio console and lets you set audio levels for each track and does similar editing techniques used in the edit window. These are basic editing features you will encounter when working with editing software. Each program may have a slightly different interface, but all digital editing software works in a similar manner. The programs record or import sound, manipulate the sound, and output the sound to a tape or digital format.

FILE FORMATS

The file formats that are used in the digital world have created some difficulties. The audio files are not compatible with each other and often have to be converted, which can cause loss of sound quality. Not all editing software can open any audio file format, so it is important to use programs that can meet your needs. Sound Forge on the PC can open multiple formats and convert them to different file formats. This makes this program invaluable to the media producer integrating audio files into Web sites. The audio file formats for Macintosh are QuickTime, AIFF (Audio Interchange File Format), SDII (Sound Designer II), System 7 Sound, and MP3. For the PC, they are WAV, RealAudio, MP3, and Windows Media. There is no one standard audio format used, but AIFF, QuickTime, MP3, and WAV appear to be the most popular formats in use today on the Internet. Another file format you may encounter is MIDI (Musical Instrumental Digital Interface), which was developed as a standard in the 1980s for electronic musical instruments and computers to talk to each other. File formats are important because, if the end user does not have the appropriate software plugin or player, he/she cannot play the audio. Many Web sites offer downloads of the players and plugins on the site to ensure that the audio can be played, but deciding which format to use is not always that simple. File format selection depends on several factors, such as end use, desired quality, and whether the audio is original or prerecorded. It is important to keep track of what audio formats can be used with the most popular Internet browsers.

The end use of the sound plays a vital role in the file format. If the audio is a musical piece, then MIDI file format provides high quality. However, if the sound is going to be downloaded to desktops, then MP3 provides decent quality and quick transfer of data, while taking up less space on the Web site. WAV file formats are a good choice if the media producer is going to burn a CD. Most end users are satisfied with MP3 quality and downloading time, but some people prefer high-end original audio. This type of audio can be found compressed into QuickTime or WAV file formats, which require some manipulation before compression. When preparing original audio for the Web, it is imperative to enhance the audio through a process called optimizing the file. In this process, one raises the overall level of the audio as much as possible without distorting the sound. This is done to decrease the loss of quality during compression, which samples the original audio but does not duplicate it exactly, thus cutting out subtle nuances of the audio in the process. Using level/Hot audio increases the quality of the audio file that is compressed. This process provides high-quality audio that takes up little space on the Web site and can be transferred quickly. The media producer also decides how to record the audio, such as at 44.1 kHz, stereo, and 16-bit rate, which is considered CD quality sound, or at 22.05 kHz, mono, 8-bit rate, which is lower-quality sound and takes up less space on a drive. There are many variations of this process when recording digital audio. The basic principle to remember is that, the higher the kilohertz, and bits numbers, selecting stereo over mono, increase the sound quality, but more space is used up on the storage device. In short, high-quality audio equals large files.

This is why compression has become so important and why QuickTime and other programs are used to shrink the size of files down. Once the audio clip is finished, it is encoded using compression software (QDesign's Music Pro 2™, QuickTime Player Pro™, Media Cleaner 5™, and so on) and is ready to be used in a Web page or in a digital presentation. There are many compression software programs available; some are higher quality than others, but all work in a similar manner. The encoding programs shrink the audio files by sampling the information discarding the redundant information while decoding programs uncompress the files so the clips can be played. The players or plugins used in the compression process are called codecs.

Basic Software Needs

The media producer has to decide on which Web authoring software to use and to keep the Web browser (such as Netscape Communicator or Internet Explorer) in mind when choosing audio for the site. There are many authoring programs to choose from, including the popular MS FrontPage™, Go Live™, and DreamWeaver. These programs are dedicated software that build Web pages and sites for the Internet. There are other programs that can be used to make Web pages or transfer format into HTML language. Word-processing and PowerPoint documents can be transferred into Web pages. The Web browsers also offer Web page-building services. Netscape Communicator has Composer, which allows people to set up a Web page from the menu bar of the browser. Many Internet providers and companies offer this service, and you can set up a Web page through AOL, Yahoo, GeoCities, and so on. Their Web page-building services are more like templates and have limited design capabilities, but you can produce a well-structured Web page for a low cost. Most of these programs allow for the integration of audio into a site.

A great deal of the sound used on the Internet is prerecorded audio clips, which people download from various Web sites and import into their own Web pages. This is why it is important to know the common file formats used, to ensure that your audio can be played by your site's end user. Many Web sites use audio that Internet users may not be able to play because they don't have the player, plugin, or updated browser, and the sites may not include links to the sites needed to help users play the audio clips. This is poor use of audio clips. Media producers get seduced by the opportunity to use sound and integrate large audio clips that either can't be played or take a long time to download. It is important to remember to use audio files that can be played by most people on the Internet and don't require high data transfer rates.

Many sites offer audio clips that can be downloaded and used on other sites. For example, MP3.com, RealGuide.com, RioPort.com, and EMusic.com all offer audio that can be downloaded and used on a Web site or played on your desktop for entertainment. Many of these sites offer free audio clips or clips for a small fee. There are thousands of audio clips available on the Internet. In fact, unless a clip is protected from being copied, almost any sound you encounter on the Internet can be copied and transferred to another site, regardless of the copyright status. Some software (such as Netscape Composer™) will let you drag and drop the clip onto your desktop and the Windows operating system allows you to right click and copy the clip. Unless the file is copy protected, it can be acquired and used in a number of ways. These audio clips can be used on Web sites, in digital presentations, in multimedia tutorials, and for personal entertainment use on a desktop or burned to a CD. If you are going to use the audio on a Web site, you must be able to play it to use popular file formats, such as QuickTime, WAV, MP3, or AIFF. Most computers produced after 1998 can play these formats and you can easily download an MP3 player. Otherwise, a link must be included to the end user to download (for free) the necessary software plugins.

The Internet browser is another factor to consider when placing sound on a Web site. Each browser recognizes certain audio files and wants viewers to use its browser over competitors. The two most used browsers, MS Internet Explorer and Netscape Communicator, recognize and play the most popular audio files. To ensure that a sound plays on a Web site, the media producer needs to open the site in each browser and play all audio clips. This procedure should be repeated at an outside computer the Web site was not developed on and ideally through a computer outside of the server at which the site will be stored. A media producer can find and correct any problems an end user may encounter when trying to access the sound on the Web site using this tip.

Now that you have an understanding of how to record, edit, and save digital sound, let's discuss the basic steps of integrating sound into a Web page using the Web editing program Go Live. See Fig. 5-7. The interface may vary a bit in different Web-editing programs, but the overall process is very similar. Step one is to acquire the sound clip. Step two is to import that sound clip into the Web page, and step three is to play the sound clip in the browser mode. The first step requires going to a Web site that has an audio clip you want to use on your Web site. The site may have a

FIGURE 5-7

A Go Live screen capture
Source: Adobe product screenshot reprinted with permission from Adobe Systems Incorporated

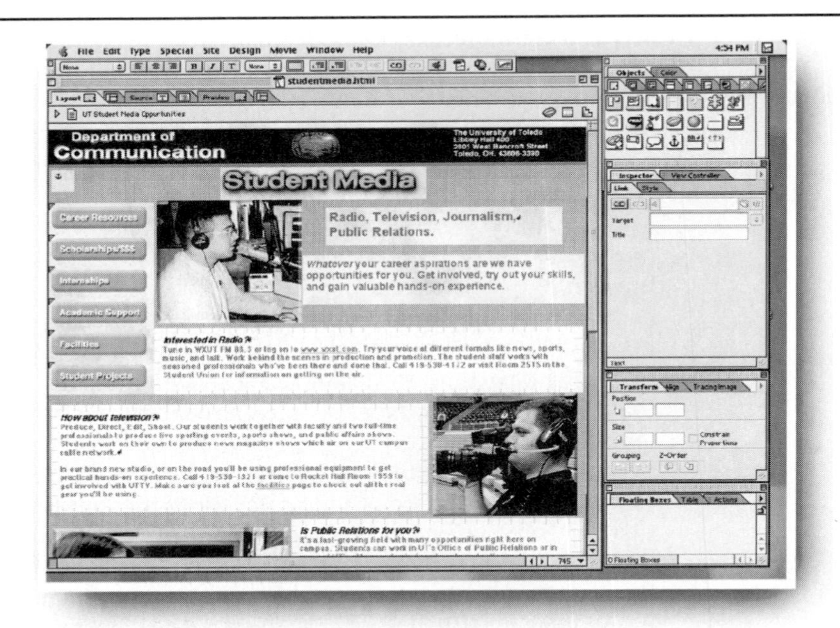

downloading button for the audio, or you can copy the file down to your computer by right clicking. Once the audio clip is stored on your computer, the next step is opening Go Live and going to your Web page under construction. The following information is just a cursory discussion of the various steps required to integrate an audio clip into a Web page using Web authoring software. The entire process may require more steps, depending on the software used in constructing a Web page or Web site.

In Go Live, go to the folder that houses the information for your Web page and open it. Then proceed to the menu bar and click on Add Files. You will see a dialog box with multiple files listed. Open the file that has your desired audio file saved in it. Select the audio file and click the Add icon. Then click on the Done icon. The audio file has now been added to the specific Web file for your Web page that you opened earlier. At this time, it is important to make sure you add the appropriate plugin for the audio file to your Web page, or the audio file will not play the sound. You accomplish this task by linking the audio file to the plugin by opening the palette folder, which houses multiple plugin software. A dialog box called the plugin inspector will open. This window allows you to choose which plugin to use for your particular audio file. In this case, the file is a wav audio clip, and the plugin you select is called audio/wav. Once the plugin type is selected, you proceed back to the plugin inspector dialog box and click on the Point and Shoot icon, which links the audio/wav file to Go Live's audio/wav plugin. You may also hide the plugin graphic, but the actual plugin needs to be present on the Web page for the audio file to play. Now select an item on the Web page to link the sound to, such as text or a graphic. If it is text you are linking the audio clip to, then the text inspector window will open. Go to Special on the menu bar and select New Link. Next, select the Point and Shoot icon and drag it to the audio file located in the file window. This will link the audio to the text. The audio file should play when you click on the text. Finally, choose the preview in a browser and click on the text to ensure that the audio clip will play. It is also important to note that you can decide how the audio is played in Go Live by having the sound loop or automatically start when the page is loaded. This is the basic process of adding sound to a Web page using a Web authoring software.

THE ART OF AUDIO COMMUNICATION

Far too often, audio is forgotten in the designing of mediated messages and yet it routinely enhances many aspects of a message. A picture of a bee without the sound of the buzz is lacking vital information in describing a bee. There are less obvious aspects to the uses of audio in the act of communicating. When to fade sound in and out or the layer of multiple sounds to create a certain environment are examples. Think of a train off in the distance, approaching, and an audio clip that slowly fades the volume up and has the sound move from the left speaker to the right speaker, and you start to get an idea just how important sound can be to a message. There is a skill to using successful audio in mediated messages; some would even call it an art. For example, Oscars and Emmys are awarded in the categories of sound design and sound editing. Sound use is more than just matching audio to pictures or watching talking heads; it is about purposely creating a mediated message in which the sound is vital but not overt. This takes a great deal of practice and work and is not done without much thought. The audio designer has become an integral part of creating mediated messages. Thus, the need for audio has increased over the years, as technology has become more pervasive in our society. Entire industries have developed to meet this need, and many people play a huge role in developing audio today. There are performers, engineers, editors, producers, and designers.

AUDIO DESIGNERS

Audio designers sometimes are performer, engineer, editor, and producer, all in one. Technology has made it possible for one person to fill many roles in producing audio. However, most designers are in charge of recording, editing, and integrating audio into various media, such as television programs, feature films, and Web sites. Designers make the decision on audio quality, file format, and where the sound will be placed in the mediated message. The audio designer must keep several elements in mind when creating audio pieces. The most important element is the audience, followed by how the audio is going to be used, and are there copyright issues. All successful designers keep the audience in mind when creating a message, and audio designers are no different. The audience is the whole reason for creating the message in the first place and must always be the key focus when designing the audio. In other words, does it make sense to use pop music when the message is being aimed at senior citizens? It may seem obvious but is often overlooked by designers who are enamored with all the bells and whistles at their disposal. A well-designed audio piece takes into consideration the various demographics of the audience, such as age, education, and culture. Another important element of audio design is the end use of the sound.

End use refers to where the audio is going to end up and how it is going to be utilized by the people who access it. Audio designers do not create sound in a vacuum; they want sound to be accessed by others. These others can be consumers, listeners, or just browsers, but the audio is designed to be used by many people. How an audio piece is going to be used affects how it is designed. If the audio is going to be used in a television program, then the audio must be broadcast-quality. However, if the end use involves the Internet, then sound enhancement, compression, and file format play a big role in the designing of the piece. It does make a difference where the sound is going to end up and how it is going to be used. This does not mean that you can't take a sound created for one media format and use it in another. In fact, new technology has allowed a great deal of transfer of audio to multiple media. The versatility has made life easier for the audio designer, but not all sound works for all situations. This is why sound composition, the placing of sound in a message, is vital to the successful use of audio in a mediated message. The bottom line is don't use sound just because you can.

Audio designers also need to deal with copyright issues when integrating a prerecorded audio clip into their media. The easiest way to avoid the issues altogether is to use original audio you have created for your media. If you use original audio, then you own the copyright and can use it anyway you desire. Remember to copyright your original audio pieces, so that others do not use your work without your permission. If you are going to use audio that was prerecorded by others, then you need to get permission to use it. Most audio on the Internet is under copyright protection.

Copyright protects writers, artists, and composers against unauthorized use of recordings. There are many different licensing rights that must be granted for you to legally use audio in another production. Licensing rights can be easily violated and must be checked before you can use an audio clip in your own media. For example, performance rights make sure the composer is paid for the use of a piece of music. Mechanical rights make sure the manufacturer is paid for the distribution of the recording. Synchronization rights make sure the composers are paid if their work is used with another medium, such as videotapes or the Internet. Finally, companies that made the physical recording hold master rights. A violation of any of these licensing agreements can cost a great deal of money. It depends on the violation, but penalties can range from a few hundred dollars to thousands of dollars per violation all the way up to a hundred thousand dollars if it is proven that the violation was intentional.

The best thing to do is to obtain permission to use the audio, and this can be done in several ways. If you want to use the audio for commercial purposes, you can contact agencies that are used as liaisons between composers and publishers. The Harry Fox agency in New York (*www.hfa.com*) is considered one of the top agencies for conducting this service. If you are interested in acquiring performance rights, you can contact the American Society of Composers, Authors and Publishers (ASCAP); Broadcast Music, Incorporated (BMI); or the Society of European Stage, Authors and Composers (SESAC). By paying these companies, you will get the right to use the original work as it was originally intended to be used, which means you cannot alter it. You can also check any CD for the publisher information and contact the publisher directly. The price of securing the rights varies, according to where you use the music, the time it will be used, who published it, who performed it, and how you're going to use it. The price can be nothing or millions of dollars. Most of the fees are in the hundreds or thousands of dollars, so securing a copyright is not a cheap endeavor.

For this reason, many media producers purchase music and sound effects libraries or pay yearly licensing fees (Fig. 5-8). There are positives and negatives to purchasing sound libraries. The positives are that you have unlimited use of the sound for a one time fee, but it can get dated quickly. Some of the negatives are that

FIGURE 5-8
A legal document

the libraries require you to purchase updated sound or allow the sound to be used in limited areas, such as just broadcast television and not the Internet. It can get expensive, but it can also save you money in the long run, depending on how much you end up using the sound library in your work. The other options a media producer has for using prerecorded audio is obtaining public domain sound. Public domain means the copyright has expired and the audio can be used by anyone for free, but be careful; the performance rights may have expired and the master rights could have been renewed. You can also try to use very old music, such as classical, in which the copyright has expired. Modern works produced after 1978 are protected for the life of the composer plus fifty years. The audio designer must keep this in mind when integrating sound into media productions. In the end analysis, it is better to pay for the rights up front than to pay the penalties for violating the law later.

THE FUTURE

The world of audio is in a continued state of flux. Every year, new technology and software are developed that make manipulating sound easier. It is now possible to make people say almost anything you want, even though they never really spoke the words. All that is needed is a digital recording of their voice. Equipment that cost a hundred thousand dollars can now be purchased for a few thousand, and everyone can become an audio producer. The field of audio technology has come a long way in 100 years from first recording sound with Thomas Edison's phonograph to the Internet and the ability to download sound onto your computer in minutes. The digital age has made many things possible that were barely imagined just a few years ago. For instance, MP3 players make it possible to record sound onto a player with its own small hard drive and never actually put it on a physical tape or disk, thus eliminating the need for physical formats. What does the future hold for audio technology?

The key to the future of audio technology is bandwidth. Bandwidth is the amount of data that can be processed or transferred from one computer to another in a given amount of time. The most commonly used distribution method has been telephone lines (28.8 kbps, 56 kbps), which have worked but are not as fast as other, new methods such as cable (DSL, T1, T3), wireless, and satellite. The faster and the larger amount of data utilized in the distribution method, the more audio it can hold and transfer to the end user. Presently, end users can listen to small audio clips in real time or whole audio programs by using streaming audio technology. As bandwidth increases, more audio on the Internet will be real-time and less time will be spent downloading sound. In the meantime, streaming audio, which allows end users to listen to the audio as it is sent to their computer, will suffice. The end user can also choose event-based audio, which takes longer to download but can be higher-quality.

Streaming Audio and Event-Based Audio

Streaming audio has made it possible to hear whole songs and radio programs on the Internet without having to wait long periods of time for all the information to download to your computer before being able to play it back. This is possible because of streaming audio software programs and formats that compress and play back the audio files. Streaming audio files deliver packets of data to a computer and play at the same time, while event-based or real-time audio files will not play until the entire files have been downloaded to your computer. MP3 may not be a pure streaming format but has had a lot to do with the success of streaming audio. It is this format that helped create the Napster phenomenon, which allowed end users around the world to download thousands of songs. The era of free copyrighted audio may have recently been limited, but the technology continues to make a huge impact on the field of audio technology. A standard CD quality audio file four-minute song could eat up 40 MB of space and take a couple of hours to download over a standard modem. By using MP3 compression the file becomes 4 MB and can be downloaded in twenty minutes. The difference is obvious, and other programs, such as RealAudio G2™, Windows Media™, and QuickTime 4™, have made creating audio files easier as well as more pervasive on the Internet. In addition, Liquifier Pro™ is another software program that is being used to stream audio and copyright protects audio on the Internet. These programs are making it possible to create, use, and protect audio on the Internet.

Event-based audio has been around for a while and consists of downloading the entire audio file before being able to listen to it. This type of audio on the Internet is typically used for small audio files but can be used for larger ones if you have the time to download the information. The playback quality is better than streaming, because there is no sound lag as the downloading is taking place because you have the entire file on your computer before you can play it. As bandwidth increases, the use for this type of audio will increase, especially for those content providers more interested in quality playback than in access speed. Content providers are becoming more interested in offering higher sound quality; thus, the demand for new technological development will continue.

Emerging Audio Technology Examples

The audio available on the Internet today is stereo-based, and audio designers are moving to a more reality-based experience. New developments, such as MPEG 4 (Moving Picture Experts Group) and VRML (Virtual Reality Markup Language) applications, are taking sound to new levels. MPEG 4 involves transmitting sound by describing it rather than compressing the audio, which will allow MP4 to be streamed, unlike MP3. This process keeps the sound quality high by not dropping important data and at the same time lets an end user download the audio over a standard modem within a decent time period. There are some glitches to be worked out, but the possibilities are endless when MP4 becomes more available to the public. VRML applications are taking sound to another dimension as well. Virtual reality developers have added 3D audio to software programs, which position sound around the listener. This is different from Dolby or surround sound; it more closely mimics the human ear, making it more interactive than in the past. The audio moves and can come from any 360-degree direction around the listener. This adds depth to the experience and will eventually move to the Internet and other media. It is currently being used in video games and flight simulators at NASA, and there are plans to use it in DVD movies.

These are just a few of the emerging audio technologies that are currently in development and will continue to change the world. The media producer's role will also continue to change as new tools offer more ways to craft a mediated message. Audio technology is one of the many tools involved in mediated messages that cannot be overlooked. Audio adds an important layer to any message transmitted over media, and understanding how to use it properly is the key to successful media design. A media producer begins this successful path by knowing good sound design, thoroughly learning the audio production process, knowing what a DAW does, being familiar with audio file formats, understanding basic software and hardware needs, taking on the role of audio designer, knowing copyright laws, and keeping an eye on emerging audio technology. Accomplished media producers must know a lot of things about a lot of stuff. Someone who is willing to be a life-long learner and never tires of the endless struggle to keep up on audio technology can successfully practice the art of audio communication. The road may seem rough, but the results are worth the trip.

BIBLIOGRAPHY AND SUGGESTED READING

Fidler, R. (1997). *MediaMorphosis*. Thousand Oaks, CA: Pine Forge Press.

Negroponte, N. (1995). *Being Digital*. New York: Vintage Books.

Pavlik, J. (1998). *New Media Technology: Cultural and Commercial Perspectives*. Boston: Allyn & Bacon.

6 Digital Video

The eyes have it

The world of visual communication has undergone some major changes in the past five years. Society has moved from the analog age to the digital age. Visuals that once took time to develop can now be recorded, rearranged, and displayed at a fraction of the time it used to take to create graphics, photographs, and video. The price of equipment used to create visual messages has also decreased, and professional-quality visuals have become commonplace. Digital video is one of the developments that has reduced the expense of producing visual communication. This technology has put experts' tools in the hands of nonprofessionals and has allowed people to become videographers, editors, directors, and producers. It is a powerful technology that makes the video camera the pencil of the new millennium. However, it is not just any pencil. This pencil can create pictures, sound, graphics, motion, and so on. The important issue of dealing with this new tool is knowing how to use it correctly to communicate. See Fig. 6-1.

Communicating in Motion

There are many ways to send information to others, such as writing a letter, showing a photograph, and speaking. Each way sends information from a source to a receiver, filtering the message through the medium. The textual message describes information. The still visual message displays the information frozen in time. The verbal message is immediate and descriptive. All three ways of communicating are used a great deal and are very important forms of sharing information, but they all lack a vital element. A moving visual message gives the viewer the feeling of being there and becoming involved in the scene. It gives the viewer immediate descriptive moving visuals that mimic real-life action. The moving visual message creates an environment that engages multiple senses of a viewer. This can aid the sender in communicating successfully with others. A medium that can copy real-life action is rich with information and should be used in a manner that optimizes its possibilities. Not all information requires moving visuals, but some messages need the use of this medium to fully understand the message. For example, sporting events are covered in newspapers and magazines but are brought to life only with the use of television and film. It is the

FIGURE 6-1
Video equipment

FIGURE 6-2
A picture of buildings

next best thing to attending sports events. There are also events that do not need moving visuals, such as a lecture. However, if you are going to use motion successfully to communicate messages it is important to understand the proper techniques. The best place to start is with the basic elements of producing moving visuals, which are composition and continuity. See Fig. 6-2.

COMPOSITION AND CONTINUITY

Shot composition and production continuity are the base for the entire production process. It is crucial to be familiar with these elements when designing digital video. There are many aspects of composition and continuity that can affect the shaping of a video message, but it is not within the scope of this book to get into this area in great detail. However, every successful media producer needs to have at least minimal knowledge of how composition and continuity play a critical role in the production process, because every time one puts a frame around something, one is saying something about it. This is the foundation of shot composition and production continuity. Each frame represents a moment in time and defines the subject in certain ways. The size of the frame has an effect on how we view the subject. For example, a wide screen video (16 x 9 aspect ratio) requires placing the subjects farther apart than does the traditional screen size (4 x 3 aspect ratio). Where the subject is placed in the frame can tell us the distance between the subject and the viewers or the subject and other objects in the frame. How the frame is in balance can also make the audience feel energetic or calm. As discussed in earlier chapters, formally balanced shots in which both sides of the frame are symmetrical create an ordered, calm environment. Informal balance creates a more dynamic environment, when opposing compositional elements are asymmetrically arranged and yet are balanced. Every inch of the frame must be arranged with the idea that one is designing a message that will evoke the desired feelings from the audience. It is also vital to understand that the shot one is framing can be affected by the shot that precedes it and the shot that comes after it.

Composition cannot be separated from continuity. The two elements work hand in hand and help create the pictorial flow that a good production must possess. Continuity is the organization of a smooth flow of events, action, or visual ideas by means of progressive changes in shots, scenes, and sequences. For a production to progress gracefully, it is necessary to change the subject in the shot, the size of the subject, or the angle of the shot. The successful media producer understands this rule and the basic elements of continuity, which are shot, scene, and sequence, and uses this knowledge to craft a visually moving message that stirs an audience to a predetermined reaction. Media producers do not randomly select close-ups, medium shots, or extreme long shots. They carefully choose the setting for each shot, the progression of each scene, and the way each sequence will fit together to move an audience in a particular direction. The media producer moves the audience in that direction by developing a script during the preproduction stage.

PREPRODUCTION

The preproduction process consists of creating a story treatment, a storyboard, and a script. See Fig. 6-3. It is during this stage of the production process that the media producer focuses on an idea and begins to mold it until a final script is approved before shooting can begin. This first step in the process is vital if time and money are going to be saved in the next two steps in the production process. Too often, amateur media producers skip this step and end up taking longer to shoot a production and trying to clean up mistakes in postproduction that could have been avoided with a thorough preproduction stage.

The first step in the preproduction process is developing a story treatment, which is a narrative description of the production. It can be as short as a paragraph or multiple pages, depending on the length of the production and the depth of the treatment. The story treatment describes the basic story line of the production and is usually written in third person present tense. An example of an opening sentence of a treatment is "We open with shots of a jet approaching Toledo and move into the cabin to a close-up of one passenger at a window seat." Story treatments use production language,

Color Plate 3-2
Can you read this text?

Closure / Continuation

form follows function

Color Plate 3-3
Lautrec's style took advantage of asymmetrical balance.

Homage to Toulouse-Lautrec's
La Gitane

Color Plate 3-1
Pollack's action painting style has no recognizable imagery. Do you see a face in there somewhere?

Homage to Jackson Pollack's
Lavender Mist

Color Plate 3-5
Three modes for mixing color in different media.

The Three Primary Color Models

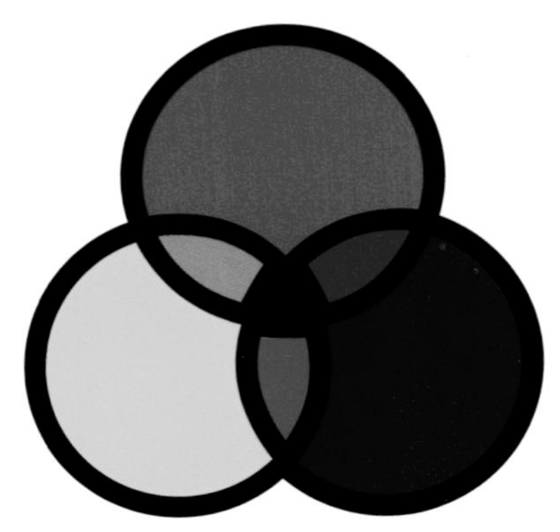

Red / yellow / blue primary colors

Red / green / blue electronic colors

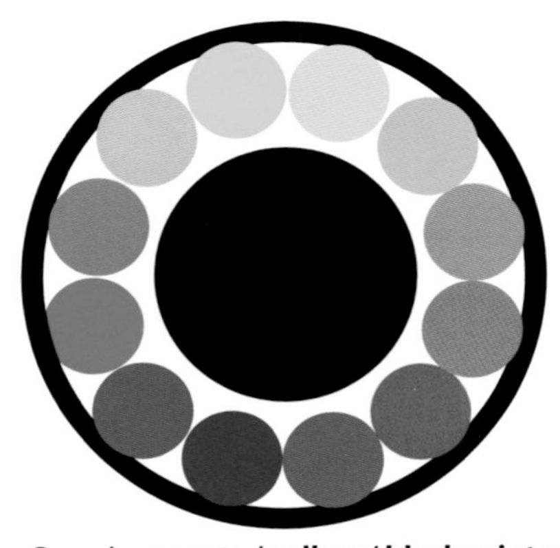

Cyan / magenta / yellow / black print colors

Directional Gestalt

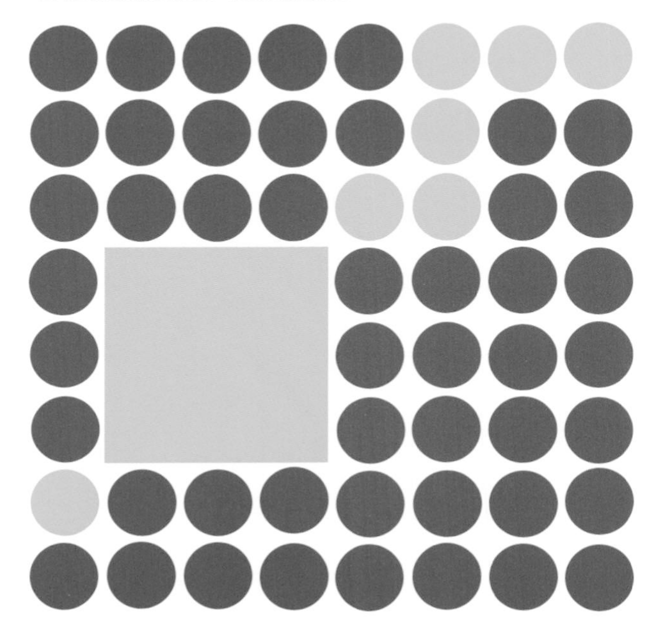

Color Plate 3-4
A designer can control the direction of a layout by using the Gestalt concept of contrast.

Color Plate 3-7
The yellow boxes are exactly the same color; however, due to simultaneous contrast, we see different yellow hues.

Simultaneous Contrast

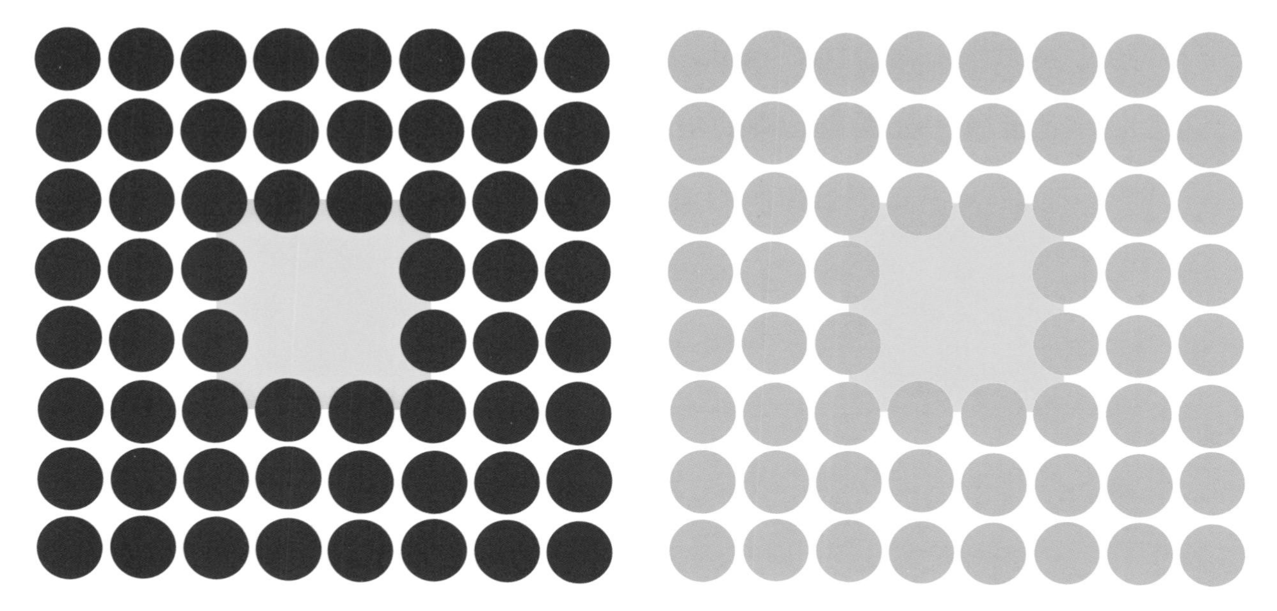

Color Plate 3-6
These are the color mixes used for drawing and painting.

Red / Yellow / Blue Color Wheel

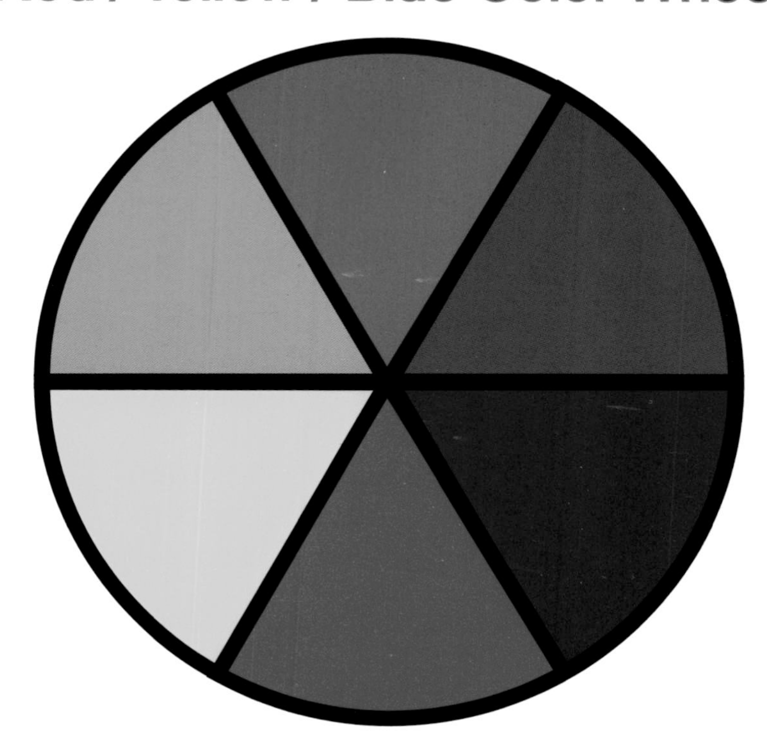

Color Plate 7-8
Ways of creating asymmetrical balance.

Color
Contrast
Shape
Texture
Position

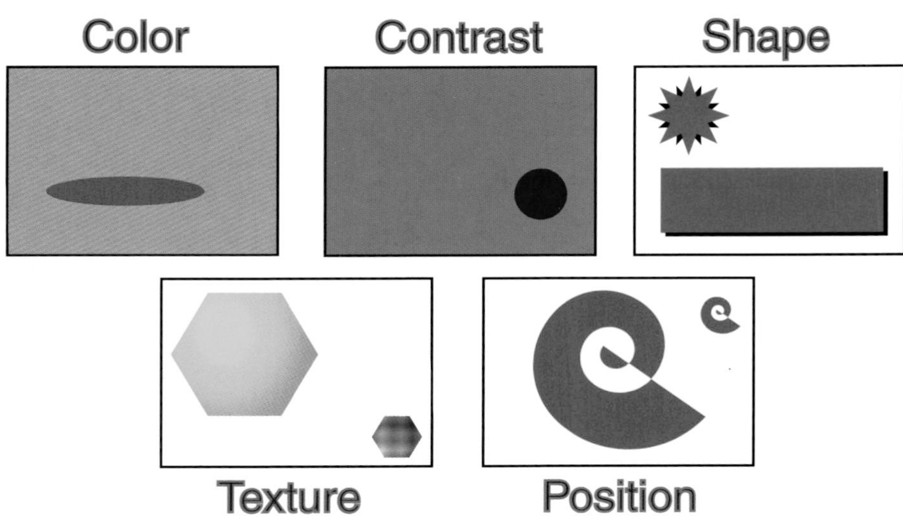

Color Plate 9-10
A sample of Apple computer's bondi blue.

The Most Popular Color

Color Plate 8-9
Text as it would be displayed in a Web page on an MS Windows platform and the same text as it would be displayed on a Macintosh platform.

Computer Monitor Displays
This is how text and color look on a Windows screen (bigger and darker).
This is how text and color look on a Macintosh screen (about 20 percent smaller and lighter).

Color Triangle

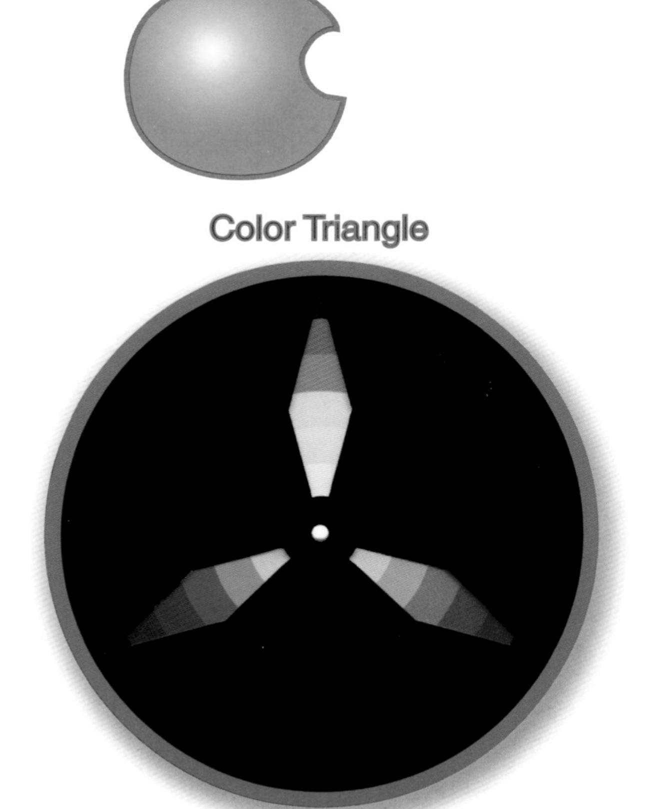

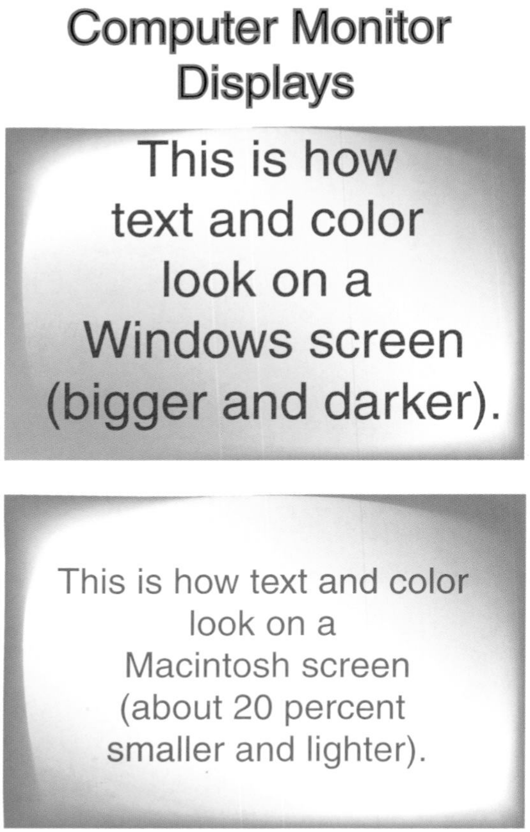

Color Plate 9-11
A triad is one of the models you can use to develop an effective color scheme.

FIGURE 6-3
A storyboard and a script

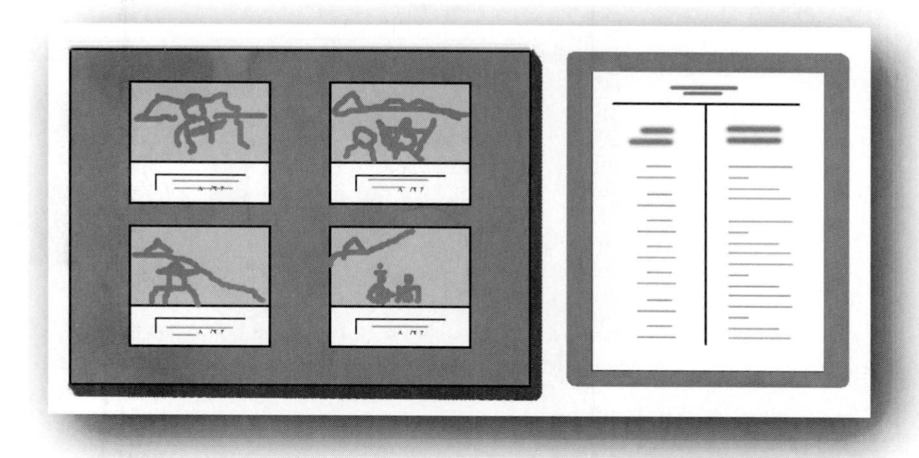

such as close-up and medium shot, as well as extremely visual descriptions. This gives the media producer an idea of how the production will look and aids in moving the process to the second step of preproduction, which is developing the storyboard.

The storyboard is a visual representation of the production, which can be changed until the desired look is acquired. Storyboards have been around since the silent era and give the producer a visual idea of what the production will look like when it is completed. It can be a large board with still photographs on it, posterboard with drawings on it, or a regular 8.5-x-11-inch paper with cards glued to it. The important element of the storyboard is that visuals are present to represent the look of the production. Storyboards can be either very complex and include every shot in a production or more of an overview, displaying only the opening shots of each scene. The photos/cards can be rearranged in any order, so that the problems can be worked out of a production before shooting begins. This is the beauty of using a storyboard when producing a video or film. Once the pictures/shots have been permanently placed on the board, visual and audio directions are added to complete the storyboard to give the production team a visual outline of the production. The type of shot and shot description, as well as who is speaking and what he/she is saying, are the visual and audio directions next to the photos/cards on the board. Many producers expect the storyboard to match the script perfectly, and some even shoot the production from the board instead of a script. An example of a storyboard can be found in Fig. 6-4.

Storyboard visuals are framed in the 4 x 3 aspect ratio, which is the screen width and length of a television screen. The visuals can be framed in 16 x 9, which is the new digital screen size. By keeping the storyboard in the same aspect ratio as the screen size, the storyboard can display a fairly accurate look of the overall production before actually shooting the production. Once this phase of the preproduction process is complete, the information on the storyboard is transferred to the script.

The script can take two forms: a full-page format, used predominately for film production, or split-page format, which is used a great deal in television production. Split-page format will be used in this book to help you better understand the production process. Split-page format places visual directions on the left side of the

FIGURE 6-4

A story board example

Source: Reprinted by permission from Microsoft Corporation

1. MS: Jill talking to class 2. CU: Student listening
1. JILL: Hello, class. 2. (NAT) Natural Sound

FIGURE 6-5

A split-page script example

	Video	Audio
:00-:05	1. MS: Jill talking to class	1. JILL: Hello, class.
:05-:09	2. CU: Student listening	2. (NAT) Natural Sound

script and audio directions on the right side of the script. The audio and visual directions are lined up across from each other, so that it is obvious what is taking place in each shot. Script times also accompany the script shots and usually give a total running time for the script. An example of a split-page format can be found in Fig. 6-5.

The example in Fig. 6-5 matches the storyboard exactly and displays how the shots flow in the script with TRT (total running times). This format aids the producer in knowing exactly what needs to be shot and how, as well as where. The preproduction process is crucial information to the media producer, who will hire the crew and rent the equipment and facilities. A thorough and streamlined preproduction stage helps keep problems to a minimum during the production and postproduction stages.

VIDEO PRODUCTION

The production stage is the second step in the production process and takes over where the preproduction stage leaves off. Using the script, the media producer, with a production crew, sets up a shooting schedule for the various shots, scenes, and sequences of the script. The production stage consists of actually shooting the script and can require a major crew or the basic crew made up of a producer, a director, a videographer, a production assistant, and an editor. The producer sets up the entire shoot and oversees the production. The director is responsible for transferring the script's ideas into

FIGURE 6-6
A video camera

the look of the overall production and controls the crew. The videographer is in charge of creating the production's visuals and of bringing the director's vision to life. The production assistant helps with audio, gripping, and any extra things needed to make the production flow smoothly. The editor is responsible for putting together the video and audio material shot during production to create a program. These crew members must be experts on the production elements in order to produce quality material.

Many elements of a production affect the quality of a properly shot script—mainly, cameras, lighting, sound, tape formats, and output. For instance, the successful media producer understands the video camera and how it works. See Fig. 6-6. It is made up of three parts: the viewfinder, the lens, and the camera itself. The viewfinder shows the videographer what is being shot and how it is framed. The lens catches the light reflecting off the subjects and sends it to the CCD (Charged Coupling Device) unit in the camera. The CCD is a computer chip that processes the light and converts it to a signal to be recorded or transmitted elsewhere. Many factors can alter an image just from the use of the type of camera (digital, analog) or the lens (telephoto or wide). A second element of the production to consider is lighting. Lighting has four purposes in production: photographic, mood, modeling, and composition. The first purpose of light is to create a video/film image. Lighting can set the mood of a production such as in a horror film. Lighting also helps make a two-dimensional format appear to be three-dimensional and gives depth to a production. In addition, lighting can be used as compositional unit to tie the production together. A good example of compositional lighting is the "millionaire" game show, which centers the light on the host and contestant when the questions get difficult.

Sound is another element that plays an important role in production. Video production is not silent for the most part and thus audio almost always accompanies the video portion of a program. Chapter 5 discusses the complex world of producing original audio, and the same elements that apply in audio production also apply to video production. The only difference between developing audio alone versus video with audio, from a production standpoint, is that the producer has to create visuals as well. This added layer of video more than doubles the complexity of producing mediated messages. Sound also adds depth to the visuals and completes the look of a production. There are entire fields that study and analyze the uses of cameras, lighting, and sound in production, so knowledge in these areas is a must for producers of mediated messages.

Another factor that affects how a production will look is the tape format used to create it. There are multiple formats available, and video production is in the process of moving from an analog system to digital formats. The analog tapes (such as Beta-Cam SP) record the information as electrical impulses on a metal or an oxide tape. The digital system records the information as a series of 0s and 1s on tapes, disks, and hard drives with little or no loss of quality as a result of transferring the information from camera to edit suite. The production process is almost the same in both analog and digital formats, except for some minor yet important details. The audio levels are different and can get distorted if you try to use analog measurement levels on a digital VU meter. The digital cameras require less light to produce quality images and are more lightweight, making them easier to use in the field.

The downside is that analog is still being used a great deal, and you will have to make a copy of your production on an analog tape for some people to view it. However, the digital system gives more versatility for the end use of a production. A digital video can be compressed or streamed and easily converted for use on computers and the Internet. The significant item to remember is how the end program or video footage will be displayed. Video productions for wide screens should have extreme long shots showing landscapes, while Internet videos require more close-ups and shots that fill the frame, because the viewing frame tends to run small. Panoramic shots do not do well on small, compressed viewers, nor do they stream well on dial-up access. In addition, low lighting does not translate well to streaming video. For this reason, it is important to keep the end product in mind when shooting a production.

The successful media producer considers all the various elements of video production when producing a program. The production stage can take place over a day or months, depending on the end length of the program, and requires coordinating multiple activities all at same time. A thirty-second commercial can take a month of field shoots or a day in the studio, taking into account the complexity of the script and the number of special effects needed. The more complex the script, the longer the production stage and the higher the production costs. It is critical that a production go well, because it costs more time and money if reshoots are required as a result of mistakes or mishaps. Once the production has been shot, the next stage in the production process is postproduction.

POSTPRODUCTION

Postproduction is the final stage of the production process; this is when the video shot during the production stage is viewed and edited. See Fig. 6-7. The postproduction process consists of transferring video footage and audio from multiple tapes or disks onto a master tape or hard drive. The master program is the end product of the entire process.

There are many ways to edit video footage. The traditional cuts-only method transfers the audio/video information directly from one tape machine to another. A switcher, character generator, and other equipment can be added to the cuts-only edit suite to create an online postproduction suite. This type of editing is called linear editing,

FIGURE 6-7
Editing equipment

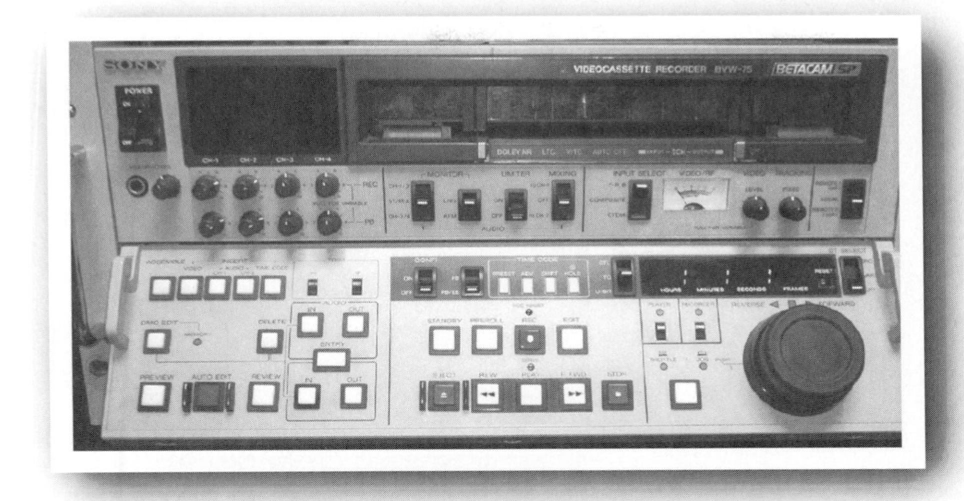

because it requires piecing the shots together, starting at the beginning and working toward the end. The current trend in editing has moved to nonlinear, or digital editing. This method allows the editor to piece the shots together spatially. The editor can drag and drop media clips in any order and start in the middle of a program, instead of the beginning. Digital editing requires less equipment and gives the editor more options to choose from than linear editing. There are many digital editing suites on the market, but the two used by most professionals are the Avid and Media 100™.

The digital video edit suite consists of a computer built specifically for editing and uses editing software. The typical hardware found in a digital suite consists of a computer with a CPU, a videotape machine, a monitor, a keyboard, a mouse, an audio console, a mini disc player, speakers, and microphones. These machines give the media producer several options to choose from when mixing various sources together. The suite can also have a DVD or other equipment that might be necessary to the production, but the basic hardware setup usually has the equipment listed. The computer can be a PC or Macintosh but needs to meet certain minimum requirements. The PC should run at least Windows 95, be a 200 MHz Pentium, with 64 MB RAM, 56 K modem or higher, 18-gigabyte disk arrays, sound card, and video card. The Macintosh should run at least OS 7.5 or newer, be a PowerPC processor, with 128 MB RAM, 18-gigabyte (150 kb = 4 minutes of video; a 4:1 ratio at broadcast quality) external hard drive or higher RAID, and 56 K modem or broadband connection. A large hard drive is required in digital editing, because video eats up a great deal of space. This is the basic requirement for doing low-end nonlinear editing. The more effects and features needed, the more updated and more powerful system you will need. See Fig. 6-8.

There are literally dozens of video editing software applications available to the public today, including the simple video recording software packaged with every Macintosh computer. The two most used software programs in digital suites are (PC) Media 100™, and (Mac) Adobe Premiere. PC-based suites also use dpsVelocity™, and Mac-based suites may also use Final Cut Pro and iMovie. Adobe Premiere and Media 100 are the

FIGURE 6-8
A Media 100 edit suite

dominant digital video editing software because they have a great deal of cross-platform versatility. These programs can mix as well as edit sound and visuals and then save the video into several file formats or master the digital video to tape, which can make the life of a media producer much easier. Let's discuss what takes place in the actual editing of video in an edit suite. The editing software that will be discussed is Media 100 and it should be considered a guide, since most editing program interfaces are similar.

The edit suite and hardware settings should be set for you to power up the system and begin editing. The editing software opens up a project with multiple windows with several options available. Once you have saved the project window, then proceed to media settings under Edit on the menu bar. See Fig. 6-9. This is where you decide the compression rate at which to store the video on the hard drive. There is a choice from 50 kbs to 300 kbs and several draft rates. The higher the compression rate, the higher the quality of video and the more space used on the hard drive. Ten seconds of video fills up 300 MB of the hard drive, and full-size full motion video requires up to 30 MB per second due to higher data rates. Obviously, storage can become an issue. This is also where you select where the media will be stored on the hard drive. It is best to separate the audio/effect and video/graphics onto different areas of the hard drive, because it will minimize glitches in playback and the computer crashes you will experience. The next step in nonlinear editing is to go to the menu bar and, under Tools, go to Edit Suite Mode and select Digitize. This will open the window that controls the tape deck and digitizes the footage. However, if you are using a camera that has firewire and an edit suite that can accept firewire, than you can transfer the digital video directly to the suite without any encoding and increase the quality level of the end product. Most digital suites use various tape decks that convert both analog and digital signals into the editing program.

In the digitizing window, you will see the video about to be digitized and can stop, play, rewind, fast-forward or search through the videotape. You can decide to store video only, audio 1 and 2, or both video and audio. There are two ways to digitize in this window. You can play the tape and click on the Digitize button and all footage will

FIGURE 6-9

An example of a digital edit suite screen

Source: "Screenshot" reprinted with permission from Media 100, Inc.

be digitized until you select the Stop button, or you can set in and out points and select the second Digitize button located next to the in and out buttons. Once a clip has been digitized, a bin window opens up and stores all the media clips.

Open up a program window by clicking on File on the menu bar and selecting New Program. The program window contains a graphic track, two video tracks, an effects track, and up to eight audio tracks. These tracks can be expanded for further manipulation of the media. You can drag media clips from the bin into the program window to build a video program. The windows you will use are the edit suite window, bin window, program window, and project window. There are many other components that you can utilize, such as the audio board window and waveform monitor. You can also change the digitizing window into the edit clip window and change the length, speed, or look of a media clip, such as by making it black-and-white. Graphics and electronic titles can also be added easily to the video program through the built-in CG (Character Generator) studio program, and audio can be imported from CDs or other sources. Once the program is complete, you can master it back to tape to be played for regular television viewing, can export it as animation, Cinepak, or a QuickTime file to another software program, or can stream it for the Internet using Media 100's Media Cleaner 5. There are several video file formats that you will encounter, and being familiar with how to use them properly can make the difference between success and failure.

FILE FORMATS

The same problems that plague the audio technology industry and its file formats also affect the video world. There is no one standard video file format. Video clips, just like audio clips, take up a great deal of storage space, even more than audio clips, and must be compressed when being exported into another application. The most popular video formats being used today are QuickTime, Video for Windows (AVI), RealVideo (RealG2), DV, and MPEG-4. There are many other formats in use, but sticking to one of these file formats should ensure easier access to playing the video clips in other applications. For example, the Media 100 allows the editor a choice of creating a QuickTime format that can be exported easily to other applications. The exported video file can be imported into a PowerPoint presentation or onto a Web page and played with the use of the QuickTime CODEC, which can be found on most computers. If one is creating original video, it is vital to save the clip in a file format that can be played back using players/viewers, such as RealVideo or QuickTime. This also becomes important when downloading prerecorded video clips to be integrated into a Web site or digital presentation. The software programs used to create a Web site or digital presentation must recognize the file format, or the video clip will not play. File formats hold a pivotal role in how video is used in various media formats, such as the Web and multimedia.

THE USE OF VIDEO IN PRESENTATIONS

As society becomes more digital, it also becomes more multimedia-oriented. Presentations that used to consist of flip charts now consist of graphics, animated charts, sound bites, and video. The ease of integrating various multimedia into presentations has vastly increased the use of video in presentations. Presenters can integrate a video clip into a PowerPoint presentation (the most used presentation software) with the click of a few buttons. See Fig. 6-10. Video clips saved or downloaded to a desktop can be inserted onto any slide and played back easily by clicking on the video or can be set to play automatically when the slide comes up. All that is required is the video clip be in a file format that is recognized by PowerPoint and the video file be saved in

FIGURE 6-10

A frame of video in a PowerPoint slide

Source: Reprinted by permission from Microsoft Corporation

a folder that accompanies the presentation it is to play in. Videos are inserted as Microsoft PowerPoint objects. However, if PowerPoint does not recognize the file format, then the Media Player is used to play it. PowerPoint saves a place mark in the presentation similar to placing a bookmark, because the video clip is rarely saved in the actual presentation because of space limitations.

Videos used for this type of display are usually kept short and simple. Most video clips are a few seconds long and use close-ups as well as basic cuts to give the video clip a smooth look. Most clips are also kept small in size on the slide and appear a bit jumpy, because the clips do not run at the full thirty frames per second required to show smooth motion. This occurs because of space limitations and is improving on a daily basis. Nonetheless, video has brought a whole new dimension to the art of public presentations. Programs such as PowerPoint also play GIF animations and sounds, making a simple presentation into a multimedia experience.

The Use of Video in Television/Films

As video is being used more in presentations, it is also being used more in television and film. The new digital system is creating higher-quality pictures with more versatility. Computer animations and graphics can be integrated easily into a program, making digital video production a very hot item in the various media industries. Movies such as *Titanic* and *Pearl Harbor* would not have been possible without the use of digital video technology. In the television industry, the portability of digital video has allowed editing on the spot and has helped productions move to multiple remote sites at lower costs to media companies. A media producer can take a laptop computer with firewire and a digital camera into the field and shoot a scene, edit it, and play it back in a matter of minutes. In the past, video production took several hours and required the aid of an entire production crew. Now, anyone with a computer and video camera can become a television producer. This technology has also allowed the media producer a choice of output methods. Once the digital production is complete, with the proper software the program can be burned to a CD or DVD or recorded in a videotape format.

The Use of Video on the Web

Web video requires a slightly different perspective than traditional uses of video. See Fig. 6-11. A successful media producer will try not to use videos produced for regular television or at the very least reedit it for the Web. Web video should have bigger titles and avoid shots with small details. Also, avoid talking head shots because of lip-syncing problems and keep the frame small for better picture quality. The length of the video clip must be short to keep the viewer's attention and open as well as close the clip with a high-quality still frame to attract the viewer to play the video. Use cuts instead of dissolves to avoid the further stuttering of the frames, and use fewer colors in the animations; it will help keep the data flowing over the connection. When mastering the video to a file format, pick AVI or MPEG and create several versions with different audio bit rates before encoding for the Web. This will ensure the best look between using music and narration. Finally, select the best digital camera possible and edit digitally. When you start with high-quality video, it is easier to maintain the quality level throughout the production process to the end product.

FIGURE 6-11

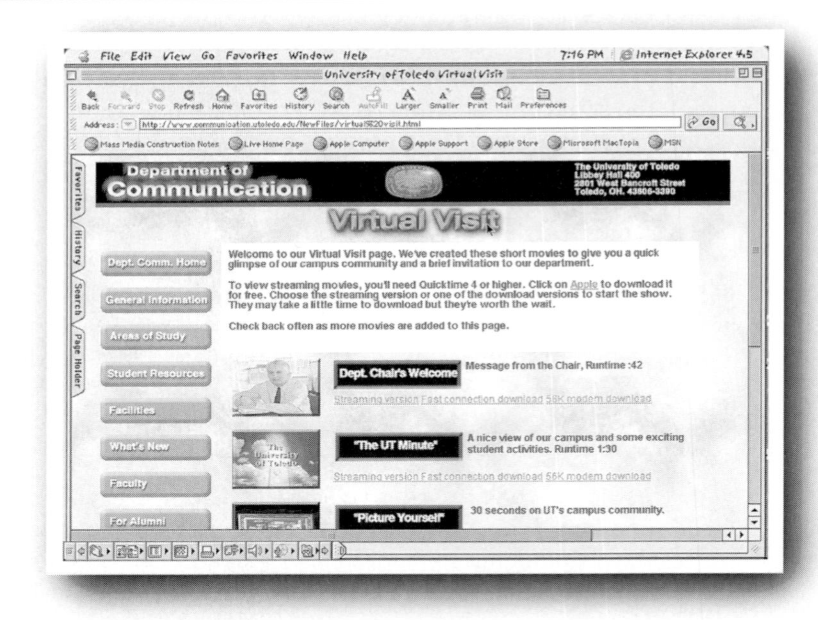

A Web screen capture with embedded video

Source: Reprinted by permission from Microsoft Corporation

Once the video is digital, it can also be integrated into Web pages. There are multiple Web authoring programs, such as Microsoft's Front Page, Macromedia's DreamWeaver, and Adobe's Go Live, which help you integrate video clips into Web pages. The process is similar to integrating video into digital presentations or sound onto a Web page. Let's discuss the basic steps of integrating video into a Web page using the Web-editing program Go Live. See Fig. 6-12. The interface may vary a bit in different Web editing programs, but the overall process is comparable. Step one is to acquire the video clip after the clip has already been saved to your computer. Step two is to bring that video clip into the Web page, and step three is to play the video clip in the browser mode. The first step requires going to a Web site that has a video clip you want to use on your Web site. The site may have a downloading button for the video, or you can copy the file down to your computer by right clicking (PC). Once the video clip is stored on your computer, the next step is opening Go Live and proceeding to your Web page under construction. The following information is just a cursory discussion of the various steps required to integrate a video clip into a Web page using Web authoring software. The entire process may require more steps, depending on the software used in constructing a Web page or Web site.

In Go Live, go to the folder that houses the information for your Web page and open it. Then proceed to the menu bar and click on Add Files. You will see a dialog box with multiple files listed. Open the file that has your desired video file saved in it. Select the video file and click the Add icon. Then click on the Done icon. The video file has now been added to the specific Web file for your Web page that you opened earlier. At this time, it is important to make sure you add the appropriate plugin for the video file to your Web page, or the video file will not play the video. You accomplish this task by linking the video file to the plugin by opening the palette folder, which houses multiple plugin software. A dialog box called the plugin inspector will open.

FIGURE 6-12

A screen capture of Adobe Go Live

Source: "Screenshot" reprinted with permission from Adobe Systems Incorporated

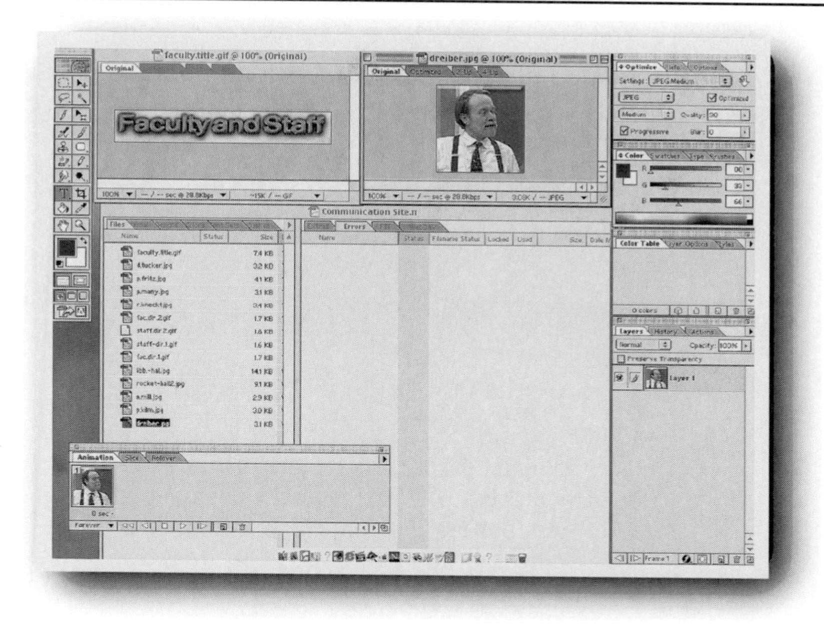

This window allows you to choose which plugin to use for your video file type by clicking on the MIME (Multipurpose Internet Mail Extension) box. In this case, the file is a video/QuickTime video clip, and the plugin you select is called video/QuickTime, considered to be one of the most used plugin software. Once the plugin type is selected, proceed back to the plugin inspector dialog box and click on the Point and Shoot icon, which links the video file to Go Live's video plugin. You may also hide the plugin graphic, but the actual plugin needs to be present on the Web page for the video file to play. Then select an item on the Web page to link the video to, such as text or a graphic. If you are linking text to the video clip, then the text inspector window will open. Go to Special on the menu bar and select New Link. Next select the Point and Shoot icon and drag it to the video file located in the file window. This will link the video to the text. The video file should play when you click on the text. Finally, proceed to the various browsers and click on the text to ensure that the video clip will play. It is also important to note that you can decide how the video is played in Go Live by having the video loop or automatically start when the page is loaded. This is the basic process of adding a real-time video clip to a Web page using a Web authoring software. When the video clip length is too long or the file size too large to download in a reasonable time period, the video can be streamed. See Fig. 6-13.

STREAMING VIDEO

Streaming video plays and downloads the video clip at the same time. The video clip must be converted to a streaming file through the use of a program such as Media Cleaner Pro™. Once a video clip has been dragged into Media Cleaner Pro, the process of hinting (compressing) the tracks begins. *Hinting* is a QuickTime term for

FIGURE 6-13
A series of video frames

adding instructions to a video clip, which tells the streaming server how to break up the video file into packets of data that are then sent over the Internet.

The process of hinting consists of a series of steps to prepare the video clip for delivery by setting the connection speed, file format, audio quality, CODEC (Sorenson), frame rate, and frame size (192 x 144). The next step in the process is to identify where the streaming video clip will be stored after it has been compressed. The streamed video clip is then uploaded to a streaming server, a server that can store and send streamed media. The final step in the process is creating a video file (reference link) on the Web page that can be linked to the video file on the server. The main difference between real-time video and streamed video is the packet data, which bundles the video data and serves them a packet at a time, instead of playing the whole video after it has been downloaded to your computer. Streaming a video clip less than two minutes long is not recommended. However, for longer clips it is essential in reducing downloading time and gives the viewer faster access to the information. This is a wonderful tool for media producers, but it can get very complicated unless the process is thoroughly understood. In addition, both real-time and streamed video clips are protected by copyright, and the media producer must receive permission to use either format of video clips.

COPYRIGHT

The media producer must obtain the permission to use copyrighted images and video created by other artists used in a new program or as stand-alone video clips. If the work was produced by the media producer, then the media producer holds the copyright. However, a great deal of video and images used in television, in film, and on the Internet are material produced and owned by others. It is the producer's responsibility to locate the company or people who hold the copyright of the media and purchase the rights to use it. Obtaining copyright permission for video is very similar to obtaining the copyrights of audio. Once the holder of the copyright is located, an agreement of how the material will be used is designed and accepted by both parties. How the material will be used has a direct effect on the cost of obtaining the copyright. One-time use costs much less than continued use in multiple media formats, such as television and the Internet. To download a video clip off the Internet, the same process must be followed. Many people make the mistake that all material on the Internet is free for the taking. In fact, even some of the media labeled "free" is not, which is an example of copyright infringement. If you are ever in doubt about the copyright, do not use the media or find the producers and obtain permission to use it. There are times when it will cost you a great deal of money, but other times it might cost you nothing except a little research time. In the end, avoiding a hefty fine for using someone else's work is worth it.

VIDEO PRODUCERS

The video producer can hold many roles, from videographer to editor to director, but the most important role a producer plays is coordinator. In today's ever changing world of technology, the video producer must stay informed about the production process, which begins with content, not the type of technology used, which is why the preproduction process is crucial to the successful use of video in communication. The video producer tells a story through visuals and always has the target audience in mind, to ensure that the message is focused. A successful video producer understands all areas of production: writing a script, choosing a digital camera, editing on a nonlinear edit suite, and using the best delivery method (television, Internet, presentation, and so on). Basically, a video producer is a Jack/Jill of all trades.

THE FUTURE

The future of digital video continues to change with each new development. Cameras and editing equipment will continue to decrease in size and grow in ability. The field of video and film will continue to merge with the field of computers. Some experts believe there will be only one field of computer technology; however, as long as there are needs for videographers, editors, writers, and producers, there will be video and film industries. Digital video will just improve the process and will allow more people to become involved in production, taking it out of the hands of the few and putting it into the hands of the masses. It is only a matter of time before you can walk into your home and the media automatically come on, creating a virtual environment you have designed. Digital video is the first step in creating the holodeck as seen on the television program *Star Trek: The Next Generation*. The technology is heading in that direction, and the field of digital video is already leading the way.

BIBLIOGRAPHY AND SUGGESTED READING

Compesi, R. (2000). *Video Field Production and Editing*. Boston: Allyn & Bacon.
Kindem, G., and R. Musburger. (2001). *Introduction to Media Production*. Boston: Focal Press.
Zettl, H. (2000). *Television Production Handbook*. Belmont, CA: Wadsworth.

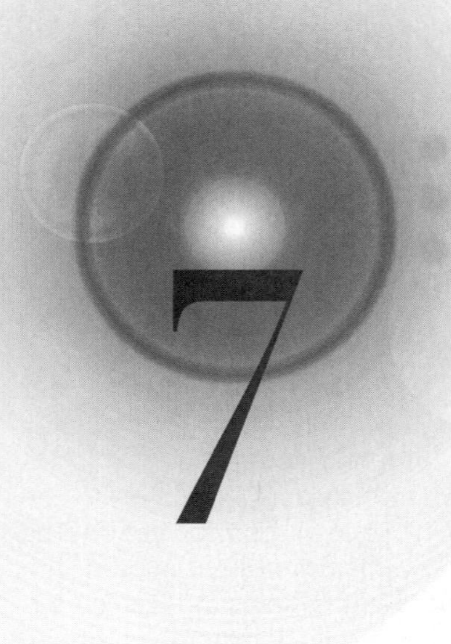

7 Digital Presentations

Can you do what I do?

Presenters have been on a quest to utilize a variety of tools to enhance public presentations. A tool that has come to dominate in recent years is the computer, and with it comes an avalanche of technologies. These technologies and other applications vary from word processing to the Internet. Today's presenter is no longer an island but, rather, is an interconnected cyber community. The explosion of this technology in society has caused many to be excited, impressed, frightened, and overwhelmed. This technology is one of the most powerful the presenter has encountered; however, it is just a tool to be used and not worshiped.

In communication, as in other fields, there is a tendency to let the tool drive the presentation instead of utilizing the tool to enhance the presentation. A situation that lends itself to being "tool-driven" is the digital presentation. These presentations are multimedia experiences. The first difficulty one runs into in understanding how to create multimedia presentations is in defining *multimedia*.

The Definition of Multimedia

The definitions of *multimedia* are many and varied, and the experts continue to argue about them. In the past, the word was literally translated into "the use of many media in one setting." Recently, the computer age has put a new spin on multimedia, and some consider it multimedia only if interaction occurs. Tay Vaughan, multimedia expert, agrees with the former definition: "multimedia is . . . a woven combination of text, graphic art, sound, animation, and video elements" (1994, p. 5), while Fred T. Hofstetter, director of the Instructional Technology Center at the University of Delaware agrees with the latter definition: "multimedia is the use of a computer to present and combine text, graphics, audio and video with links and tools that let the user navigate, interact, create, and communicate" (1995, p. 3). Vaughan is more precise in his definition and believes that there are different types of multimedia. He defines *interactive multimedia* as the combination of media and user interaction, while multimedia itself is just the combination of media.

For the purposes of this chapter, *multimedia* will be defined as the combination of many media in one setting. *Presentation* will be defined as a descriptive or persuasive account. Thus, "the integration of multimedia into the rites of presentation combines contemporary technology with the timeless presentation objective: persuasion" (Lindstrom, 1994, p. 23). This chapter discusses the use of multimedia in business presentations. It focuses on utilizing multimedia as a tool to enhance presentations and introduce multimedia to the novice.

Multimedia presentations are not new. In fact, presenters have been using multiple media since the cave people used drawings on cave walls to tell stories. The difference in today's multimedia presentations is that digital technology allows the presenter to access information in any order. The presentation does not have to be linear but can be rearranged, edited, and manipulated in a moment's notice. For example, a speaker is about to present information on a current space shuttle launch that revealed a new discovery an hour before his/her presentation. In this digital age of presentations, the speaker can edit the presentation with this new information a minute before it is due. This was not possible before the development of digital technology. The development of digital technology has brought about amazing changes in how information is presented and has aided in the information explosion.

This explosion has caused information overload, which continues to produce more information with less time to assimilate it. Multimedia presentations become more vital by providing a great deal of information in a short time span. Multimedia directly addresses the problem of information overload and overtaxed attention spans through the use of multisensory engagement. It displays information in a dynamic and time-based fashion so as to encourage feedback and interaction from the audience. Still, producers of these types of presentations need to train themselves to understand the principles and techniques of integrating media for better communication. Multimedia, by itself, does not make a good presentation. All the bells and whistles may hold the audience's attention for a while, but only a valid presentation that is well organized and well prepared gets through to the audience. Therefore, it is vital to discuss the foundation of proper presentation style.

PRESENTATION STYLE

The organization of a presentation should include an introduction, a body, and a conclusion. Also, the importance of audience analysis cannot be overstated. It is vital the presenter understand the audience members by analyzing their education, occupation, age, gender, socioeconomic background, and so on. This aids the speaker in molding the presentation to that audience and increases audience retention of the message. The digital presenter must determine the goals and objectives of a presentation, just as the nondigital presenter had to in the past. For example, a main goal of a presentation can be to educate workers about multimedia. The objective of the presentation is to teach the workers how to use multimedia for training. The content of the presentation defines the facts and figures the presenter plans to communicate, and the concept of the presentation encompasses how the information will be displayed. It is at the concept stage that the presenter selects the best technology to communicate the message.

When the presentation is to take place has a distinct impact on the overall success of the presentation. The time of day can greatly alter audience mood. A presentation that occurs right before lunch or after dinner can tire an audience and reduce their attention span. The speaker must keep this in mind and refine the presentation accordingly to keep the audience involved. Business presentations are notorious for occurring during or after meals; therefore, it is imperative that the digital presenter be prepared for this when creating a presentation, since a great deal of digital presentations happen in darkened rooms. In addition, a presenter should never ask an audience to sit for longer than ninety minutes. After that, the audience tends to lose interest and needs a break to refresh their energy levels. The digital presenter should also plan double the time he/she thinks is needed to produce a multimedia presentation. In other words, a digital presenter requires more time to create a presentation than a nondigital presenter, and this should be factored into the proposed date of the presentation. Scanning photographs, designing visuals, and producing videos increase the time needed to develop multimedia presentations.

Where a presentation occurs is just as vital as when it occurs. The audience size, the room, and the environment play a major role in the creation of a digital presentation. The audience size and room affect the size of the screens or number of monitors required to present a digital presentation. The environment of a room, which includes the lighting and sound acoustics, can make or break a digital presentation. For example, a room without a dimming panel gives the digital presenter with an LCD projector only one option—turn off all the lights. This is far from a perfect speaking environment. The audience cannot take notes and, if the multimedia is designed to be combined with a speaker, the energy of the presentation might be lost. A good number of digital presentations are designed to include a presenter but are displayed as stand-alone units. These presentations are enhanced through the use of gestures, which would be lost in a blacked-out room.

GESTURES

The successful delivery of multimedia presentations includes the proper use of gestures, facial expressions, eye contact, movement, and vocal quality. The digital presenter should use natural gestures, not choreographed ones. Too often, digital presenters try to put on a show, which is sure to fail if they are not performers. It is best to stick to basic speech principles. When using digital technology, the presenter must be completely at ease using the mouse and keyboard. This should help limit nervous gestures of tapping a pen or stuffing hands in pockets. The most common mistake made by digital presenters is being tied to the mouse and/or keyboard. These speakers have a tendency to stay close to the computer and sometimes never remove their hand from the mouse during the entire presentation. The effect is a boring presenter. The use of facial expressions is vital to a presentation. A smile shows openness and motivation. Eye contact is also of the utmost importance, because it creates a bond with the audience and keeps their attention. The common mistake in this area is looking at the computer monitor or screen too much.

Posture and movement affect the presentation by displaying to the audience the presenter's energy and authority or by diverting attention from the presentation. For example, a presenter glued to a computer can put an audience to sleep in minutes.

A digital presenter should never stay behind the computer but should move about the room, returning to the computer only when necessary.

Voice and vocal quality are also important because many speakers cannot be heard in the back of rooms. These presenters also speak in monotone voices, which can destroy a presentation. A clear, strong voice that avoids "ums," "ers," and "uhs" is best. The tone of a presentation is often augmented by the tone of the speaker. This factor is as important to a multimedia presentation as is the quality of the technology.

INTERACTION

Every good presentation offers the opportunity for interaction. The audience responds to the presenter, and the presenter responds to feedback from the audience. This interaction aids in making new things familiar and familiar things new. For example, if audience members have never heard of multimedia, the presentation on multimedia must make this topic familiar and commonplace. This will help the audience remember the topic and curb technological phobias. On the other hand, if the audience is familiar with multimedia, the presentation must make the topic new in their eyes to increase interest and not bore them. It is through this interaction that technology can enhance the effect of the presentation.

> Our concern for the effective use of instructional technology is based on the principle that information and ideas can best be communicated by a proper combination of materials and methods rather than by the printed and spoken word. (In Vlcek & Wiman, 1989, p. 73)

The first challenge for the presenter is to discover the correct materials and methods for the unlimited speaking environments encountered by a speaker. The second challenge is to evaluate the materials and methods selected and measure the effectiveness of their implementation. In order to accomplish these tasks, the evaluation of the material must follow a logical pattern to secure accuracy. Thus, the digital presenter is in the business of utilizing symbol systems to create messages and transfer information. It is through the accurate use of these symbol systems, such as multimedia, that a presenter transcends the traditional limitations of a lone speaker.

MULTIMEDIA ELEMENTS

Human beings are multimedia communicators. Several elements make up the design of any good multimedia presentation. These elements include written language (text)/narration, illustrations/photographs, charts/graphs/diagrams, sound, and motion (video/animation).

Written words are descriptive, detailed, and direct. Text can be powerful when skillfully employed, but it cannot replace the speaker's gestures, vocal quality, and so on. For example, a resume can get one a job interview, but it's the interview that gets one the job. Rarely are people given a job without an interview, because resumes are one-dimensional, while interviews provide more detailed information about the job candidate. Narration can be informative as well as expressive. The presenter should use words carefully, because misunderstandings are common. Illustrations and photographs can tell stories, grab attention, and describe people, places, and things, but they are frozen in time and locked in silence. For example, a photograph of a busy assembly line cannot convey the hum of the machines that need adjustment.

A chart, graph, or diagram adds a spatial dimension to printed data but can only suggest the element of time using dates and scales. A three-dimensional chart, graph, or diagram and creative thematic elements can alert a complacent audience. Sound and motion engage the senses on multiple levels and create multimedia with the addition of one or both to a presentation. These elements are closer to the multimedia experience of life. The combination of these various elements in a presentation create multimedia. The mere combinations of these elements do not guarantee a well-designed multimedia presentation, however. There are "proper" ways to design multimedia presentations.

SCREEN DESIGN

The design of a screen that is going to be projected for a presentation is different from the design of interactive multimedia and yet follows the same basic design principles. These principles are made up of layout, balance, visual clarity, backgrounds/textures, thematic design, and color/perception.

Good layout means proper use of screen space. Designers often use the rule of thirds to keep layouts visually interesting (Lindstrom, 1994, p. 201). The rule of thirds consists of dividing the screen into thirds vertically and horizontally. The intersections of these lines are called the four centers of interest. These points are the most visually interesting areas on the screen and where important information should be placed. See Fig. 7-1.

Compositional balance is also imperative to good design. There are two kinds of balance: formal and informal. Formal balance occurs when screen elements are symmetrical. When opposing elements are asymmetrically arranged yet are balanced, the balance is called informal. The latter is more visually interesting in nature. See Fig. 7-2.

The goal of good screen design is visual clarity. In other words, too much information on one screen is confusing. Powerful design guides the viewer's eyes to important areas on the screen. For example, a cartoon of a man pointing grabs the viewer's attention and then guides the viewer toward important text. Good visual clarity allows the

FIGURE 7-1

An example of the use of the rule of thirds

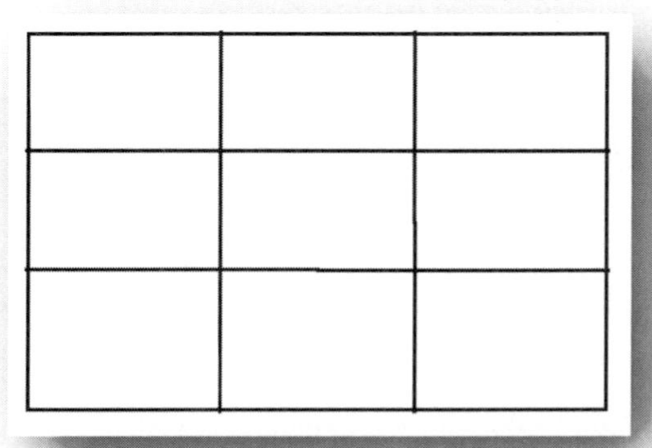

FIGURE 7-2
Balance

Formal balance   Informal balance

viewer to move easily through the screen. Backgrounds serve as the foundation for the other screen elements and must be used to underscore the message, not diffuse it.

The use of textures can aid backgrounds in adding depth to a presentation. For example, marble conveys a cool and solid feel, which translates as classical feeling (Lindstrom, 1994, p. 207).

A thematic design of these elements can bind a presentation together and draw the audience into the theme, and, as stated in Chapter 3, color selection is also vital to a presentation and the most subjective design element. A basic factor to keep in mind is that bright colors draw attention and muted colors blend into the background. Color perceptions vary but can be shared. Remember, red can mean warmth, life, love or pain, death, and Satan. It depends on the different screen elements used to create a message.

TEXT

The use of text, illustrations, and charts in a presentation must take into account the various design elements. In addition, the careful use of text is critical because words are easily misinterpreted. Text dominates most multimedia presentations; thus, color, font, and size must be manipulated to keep interest without boring or confusing. Too often, presenters design the text for desktop publishing purposes, not multimedia presentations. The result is hard-to-see text with very low readability. Text designed to be read 2 feet away on a computer screen is not read easily 20 feet away on a projected screen. Therefore, a presenter should try to follow a basic set of rules when designing the text for a multimedia business presentation. First, the text size and font type depend on how the presentation will be displayed. According to Putman (1997),

> Print out the fonts you want to use on a piece of 8.5"x11" paper. Draw a 6"x8" box around the fonts, and hold them at arm's length. This is how large those fonts will appear on a 12-foot wide screen at a distance of 30 feet. (p. 54)

In other words, the type size has to be much larger in multimedia presentations than in desktop publishing. Type size should increase as the distance and screen size increase:

For example, text on an overhead projector in a classroom of 25 students needs to be at least 20 pts. or higher.

Thus, text displayed on a rear-projected screen in an auditorium that seats 500 people requires a larger type size. The font type depends on color and overall screen design. Keeping it simple is another good tip. There are many different types of fonts, from *machine to script*. One font looks like a `typewriter,` while another appears to be ***calligraphy***. The most important thing to remember when choosing fonts is the readability factor.

COLOR

Color plays a major role in the design of multimedia presentations and is usually overlooked. Multimedia presenters often manipulate color without regard to the effects on the audience. Many presentation software programs provide guides to color templates. These templates can be altered, but sometimes with poor results. The problem that occurs is that the designer changes a color in the template with a color close in gray scale value to the other template colors. This creates lower contrast and makes it more difficult for people to view. As Itten observed, "A color is always to be seen in relation to its surroundings" (1970, p. 91). This is why it is vital that the colors selected do not share the same gray scale value. Gray scale value is the various shades of gray from light to dark. The closer colors are in gray scale value, the less contrast and the more difficult it becomes to read the text. Colors should enhance the message, not make it more difficult to deliver the information to the audience. The templates are designed with color harmony in mind. As Wong states, "Color harmony is best defined as successful color combinations, whether these please the eye by using analogous colors or excite the eye with contrasts" (1987, p. 51). If presenters are not sure about color selection, then they should stick to the templates suggested by the presentation software. Presenters with basic color information can create unique multimedia presentations.

There are three basic uses of color in multimedia presentations: to identify, to contrast, and to highlight. Color can be used as a theme to identify the main topic or used to contrast the main topic with less important topics. It is also used to highlight information on the screen to draw attention to the information on that area of the screen; color selection of text, outlines, backgrounds, shadow, is crucial to the success or failure of a presentation.

There are certain guides to remember when utilizing color in digital presentations. The colors that seem to project forward and dominate the screen are reds, yellows, and oranges, while purples, blues, and greens appear to recede into the background. The novice digital presenter would be wise to limit a color palette to no more than six colors. This usually promotes successful color solutions. Brightly colored shapes look larger than darker ones. This is an important tip to recall when using charts and illustrations. For example, yellow bars in a chart can be used on the statistics the multimedia presenter wants noticed, while blue bars display less significant information. As mentioned earlier, colors also contain emotional associations, which can affect viewers' perceptions of digital presentations. Red used on a screen displaying wedding information creates an association with love and warmth; however, red used in a screen displaying crime statistics creates an association with pain and evil. In order to produce effective digital presentations, presenters need to be aware of these various associations. Color is one more layer of a multimedia presentation that aids in transferring information to the audience.

ILLUSTRATIONS

Illustrations, such as clip art, drawings, and diagrams, are yet another layer of a multimedia presentation used to enhance the information being provided. Nonetheless, illustrations are frequently used only to grab attention and thus confuse the message. There has been an overuse of clip art in digital presentations. This occurs because software programs are providing more and more clip art galleries. Presenters feel compelled to use clip art on every screen. However, good screen design requires the use of illustrations only when necessary. For example, suppose a presenter needs to explain a change in a park facility. A drawing of the facility will clarify the information to the audience. Illustrations can also be used to produce themes, concepts, and symbols. Clip art can be a vital tool to the digital presenter in these areas.

Most clip art galleries are a part of software programs that allow the manipulation of the clip art (Fig. 7-3). In other words, the vector drawings can be ungrouped and rearranged to fit a particular presentation. This versatility can be good and bad for the business presenter. It's good when a presenter needs only a part of the clip art, such as a podium without the speaker, but it's bad when a presenter abuses this tool—for example, by resizing the clip art and not keeping proportions in mind. The result is a skewed illustration that detracts the viewers' attention from the main information of the presentation.

CHARTS AND DIAGRAMS

Charts (see Fig. 7-4) and diagrams are ideal for data visualization but can be overused and bore an audience. For example, suppose a presenter needs to explain a change in a floor plan. A diagram of the floor plan will clarify the information to the audience.

FIGURE 7-3
Manipulation of ClipArt

FIGURE 7-4
A chart

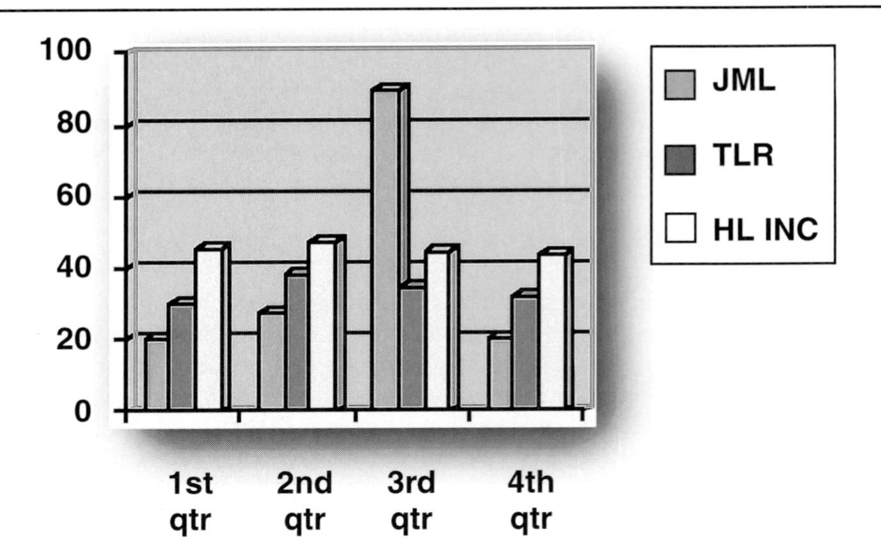

A presentation with a series of diagrams of floor plans will clarify but can also put viewers to sleep.

It is vital to use color and motion in charts and diagrams. The use of animation with charts has proven highly effective in keeping viewers' attention. A piece of a pie chart moving from the pie to the top of the screen can energize a presentation and aid in explaining the information. Each piece of pie can move to a different part of the screen, which further explains the data. The data are visualized and can be compared easily. If animation is not in a presenter's budget or time schedule, the creative use of color in charts and diagrams can also aid in the visualization. Bright, vibrant colors give life to a chart and highlight areas of a diagram; however, the overuse of bright colors can be obnoxious.

The presenter should always keep in mind that charts and diagrams are used to enhance the message, not to confuse it. Digital business presenters have a tendency to overuse charts and need to rethink how information can be displayed in a variety of ways. The right balance of illustrations, charts, diagrams, and other graphics can make or break a presentation.

Photographs and Scanned Images

Photographs and scanned images are visually rich and can grab the viewer's attention. These images bring reality into a presentation and provide detailed information. These types of graphics are very powerful and should be placed at a visually interesting point on the screen. Photographs and scanned images draw the viewer to a point on the screen and then guide the viewer through the screen. For example, a photograph of a model grabs the viewer's attention, and the light-colored text on a dark background guides the viewer to a description of a model's life. The photograph enhances the textual information. All screen elements should complement and enhance the message of the presentation.

FIGURE 7-5
Editing out a person

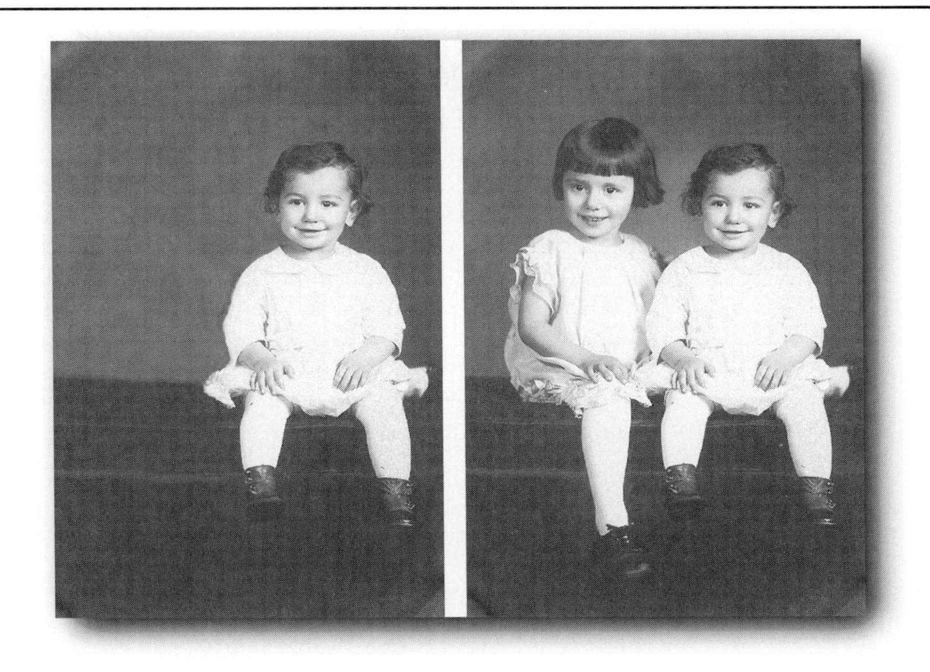

Integrating photographs into multimedia presentations has become quite easy because of programs such as Adobe PhotoShop and scanners. Adobe PhotoShop allows the presenter to customize images for a presentation. A presenter can erase people from a needed photograph (see Fig. 7-5) or place a subject in another environment vital to the message of the presentation. Several photographs can be made into one image, and one image can be made into many images. The versatility of this program is amazing, and it has brought multimedia presentations to a new era. Nonetheless, it can be overused and images are often overedited, which tends to confuse viewers. A good rule to stick to is to keep it simple.

BALANCE

The placement of graphics and text on a screen is a delicate balance between screen elements. As discussed in earlier chapters, there are two types of balance in any design. Symmetrical balance is the simplest type and the most recognizable. This type gives equal weight to both sides of the central vertical axis of the overall design. The design can be folded in half, and each half mirrors the other. The second type of balance is asymmetrical, balance achieved through the use of dissimilar objects.

There are several ways to create asymmetrical balance in a design. One way is to use color, which has been proven to attract viewers' eyes (Lauer, 1985, p. 47). For example, a small area of bright color can balance a larger area of neutral color. Another way to create balance in screen design is through contrast. A contrast of dark and light attracts viewers' eyes through the value difference. For instance, "a darker, smaller element is visually equal to a lighter, larger one" (Lauer, 1985, p. 47). A third method of creating balance is by using shape. The difference in shapes on the screen draws the viewer's attention. For example, "a small, complicated shape is balanced by a larger, more stable shape" (Lauer, 1985, p. 47). Another mode of creating screen balance is through texture. Visuals that contain various textures are more visually interesting than smooth surfaces. As Lauer observed, "a small, textured shape can balance a larger, untextured one" (1985, p. 47). A fifth way of creating asymmetrical balance is by position. Placing a larger object near the middle of a design can balance a smaller object placed toward the outer edge (Lauer, 1985, p. 47). Finally, balance by eye direction is used a great deal in design. According to Lauer, "a single small element can be as important as many larger ones if it is made the focal point of the design" (1985, p. 47). For example, many large arrows pointing at a small graphic of the sun draws the viewer's eyes to the sun, not the arrows. See Color Plate 7-8 in color insert. These methods of asymmetrical balance are often used together. They are layered to produce an interesting design that aids viewers in gleaning information.

SOUND DESIGN

Five factors must be considered when utilizing sound in a multimedia presentation: (1) heightened interest/impact, (2) better retention, (3) image enhancement, (4) comic relief, and (5) perceived value. Music and sound can keep an audience tuned in and focused by heightening interest while creating an impact—for example, adding a fanfare sound effect to a slide with good news or adding a foghorn sound effect for bad news. Sound should also be used to aid the audience in better retaining the message. Sound can connect the feeling of music with the feeling the presenter wants the message to convey; in many cases, that is how the message will be remembered—for example, a jingle from a commercial. The right choice of musical enhancements can help establish the presenter in the mind of the audience, as just about any image the presenter desires. For instance, the presenter may wish to be thought of as cool, laid back, classical, or businesslike. The type of music affects the mood of the audience and how the presenter is identified. Comic relief makes a presentation more effective by giving the audience a chance to breathe and wake up. It may take nothing more than an offbeat sound effect or a laugh track, but it must be appropriate for the audience. Humor can make or break a presentation and must be used carefully. Finally, perceived value is the last factor to consider when utilizing sound. The higher the quality of the sound elements, the more positive will be the response to the message. For example, MIT (Massachusetts Institute of Technology) tested subjects on their perceptions of the quality of TV sets with identical pictures, but one set had poor sound. The subjects thought the set with better sound had better picture quality.

A digital presenter needs to use high-quality clip music and sound effects. Speakers should also avoid using low-cost equipment for use in presentations. This may seem obvious, but many presenters work on a budget, and it is an easy place to reduce costs. Presenters ought to look for samplers and free demos to get a feel for the different types of sound that are available. This aids the digital presenter by reducing the production costs of a multimedia presentation. It is imperative that presenters not use copyrighted music, unless they have paid for the rights. Furthermore, public domain audio is available for use and can be purchased at reasonable prices.

Multimedia has a tendency to overkill the use of sound in a presentation. Sound should be used to enhance or draw attention to the presentation, but the overuse of it is irritating and can turn off an audience. A basic guide to decent sound design is to use sound to evoke reactions from the audience without confusing them. Digital technology has allowed the multimedia presenter to become a conductor of an orchestra. By the touch of a button, opera, rock-n-roll, the blues, narration, and more can become a part of a presentation. For example, a CEO can make a statement to employees across the country without leaving the office through a digital presentation with narration from the CEO. Digital presentations use sound in a very cost-effective way. Sound adds texture to visuals and can entertain an audience as no other design element can. It should be treated with the respect it deserves, not as an afterthought.

Too often, presenters record the audio for a presentation at a lower quality to save disk or hard drive space. The result is a noisy popping sound, which can destroy an otherwise great presentation. To ensure this does not occur, presenters must record audio at the highest possible quality. A good level at which to record and play back is 22 kHz. Examples of other sampling levels are 44 kHz for a CD and 4 to 8 kHz for a speech. It is important to understand that higher quality also means that more memory is required. Presenters should be aware that storing digital audio can use up a great deal of space. For example, one minute from a standard music CD at the same quality level takes up about 10.6 MB of memory (Lindstrom, 1994, p. 296). This is one more reason to use sound only when it is necessary. Finally, a presenter should know the type of hardware to be encountered in various speaking situations. It is quite possible that presenters will have only the most basic equipment with which to work. A presentation designed to use basic equipment can work in most situations without technical problems. This requires lower-quality sampling levels, which is not preferable but is manageable.

MOTION DESIGN

In multimedia presentations, video has the power to attract attention at trade shows, to record testimonials, to take customers anywhere in the world, and to overcome barriers of culture, language, and illiteracy. It can bring new employees up to speed on the use of complicated machinery and lend credibility to the presenter, the message, and the organization. The use of video in multimedia presentations can tell a story in a linear or nonlinear fashion. Nonetheless, the presenter should use video only when it is clearly the best way to make the point. There are four main reasons to use video: (1) to describe motion, (2) to demonstrate procedure, (3) to convey emotions, and (4) to stimulate situations. If motion is an integral part of the message, then video is an obvious choice. Video can also be used for the accurate demonstration of tasks—for example, the operation of a tractor. This medium excels at displaying emotional and psychological interactions. For instance, it can aid in product identification. Video can copy real-world situations or build hypothetical events (Lindstrom, 1994, p. 321). A video segment using actors to show managers how to deal with diversity issues is a good example.

The producing of video for multimedia presentations must take into account how the video is going to be captured and played back. The window size, which represents the number of pixels displayed horizontally and vertically, must be selected. The frame rate, which varies from a slow 10 fps (frames per second) to full-motion 30 fps, must be selected. The image quality, which is the amount or bit depth of digital information that the codec (compression/decompression) device captures each frame, must be selected. The data transfer rate, which creates a smoother motion effect with higher speeds, must also be selected. Thus, a larger video window size stored at a higher frame rate with a faster data transfer rate and at a higher bit depth image quality requires more memory.

The placement of the window depends on the size and the placement of text or other design elements. The window is usually placed in the center of the screen, which is the weakest compositional area on the slide. It should be set off center slightly or placed near the top third of the screen to draw attention to it. See Fig. 7-6. Text should be simple and concise. Any other graphics on the slide should enhance the use of video and not divert the audience's attention by causing confusion at what to view. The standard window size for presentation is 160 x 120 pixels, while the usual frame rate is 15 frames per second. This displays a small window with somewhat jumpy but acceptable motion. The standard window size and frame rate continue to increase as processing speed, RAM storage, and hard drive storage increase. However these basic tips on where to place the video window should still be a good guide.

Shooting video for multimedia presentations is different than shooting for a regular video production. For instance, the shots in multimedia must be cropped. Close attention must be paid to framing the shot by filling the frame with the subject. Since the window size is usually small, added space around the subject makes the subject appear even smaller. The video's movement must be slow and steady, because compression schemes are most effective when sequential frames are similar. Too much movement in the video will make it appear more jumpy and minimize the overall quality. The lighting for multimedia video must be direct and even. Hot spots and dark shadows should be avoided, because they can wash out or easily hide the subject. The color schemes utilized must be subtle, to allow for more tonal information to be captured without overloading the available bandwidth. Finally, high-quality audio equipment must be used, because this is the most overlooked component of multimedia. As discussed earlier, sound quality affects the audience's perception of picture quality.

FIGURE 7-6
Composition

PRESENTATION SOFTWARE

According to Vaughan, "presentation software might, indeed, be considered authoring software, because the publishers of today's products are making them more and more multimedia-capable" (1994, p. 138). These programs are widely used and aid in preparing speakers for more complex authoring programs. In fact, most digital presentations use programs such as MS PowerPoint. These popular programs consist of a series of layout and color templates that can integrate graphics, sound, and video. This software also allows the presenter to create original layouts and color templates. Photographs, scanned images, audio, and video previously digitized can be placed within any slide. A presenter can use effects such as dissolves, wipes, and fades before, after, and between slides. There are even basic animation effects available. MS PowerPoint can integrate animation, but only through an animation software program. Presentation software programs also offer a variety of hard copy, including outlines, slides/notes, slides, and a slide per page. These programs are user friendly and do not require a great deal of time to learn, but they also have a great deal of versatility.

A presenter can easily convert ready-made reports into a presentation and vice versa. The common mistake made in using these programs is slapping together slides of charts and text without regard to screen design or the audience. For this reason, presenters often turn to higher-end authoring programs for more bells and whistles. Authoring programs such as Macromedia Director™, Authorware™, and Toolbook™ can create multimedia presentations that are more interactive, but these programs are more difficult to learn. Authoring programs require a great deal of time to prepare but offer the ability to create stand-alone presentations. In other words, the viewer can activate the presentation and controls the pace, as well as the direction. Presenters using these programs encounter a myriad of technical problems, and they require higher-end equipment to display. A presenter should work up to these programs and use them when it is appropriate for the speaking environment. Presentation software programs, such as MS PowerPoint, more than suffice in most speaking environments.

PRESENTATION HARDWARE

The basic presentation hardware setup includes a computer, an LCD projector, and a screen or monitors. Computers used in digital presentations can vary from portable desktops to laptops. They can be Pentiums or Mac PowerPCs. The more graphics, sound, and video used in a presentation, the larger the memory space is required. The basic computer specifications required for most digital business presentations are 133 MHz, 16 MB RAM, 256 KB Cache, 1GB hard drive, integrated 16-bit stereo sound, 10X CD-ROM, and 2 MB EDO ViRGE 3D video. This setup should work for most presentations; however, as graphics, audio, video, and quality increase in presentations, so do the specifications. Higher processing speeds, more RAM, and larger hard drives are becoming necessary for multimedia presentations. A digital presenter will also encounter LCD panels and projectors used to project the presentation onto a screen. The panels are used with a powerful overhead projector in classroomlike settings. The LCD projectors are used in conference room settings. This equipment requires the use of a screen to be displayed properly. Rear-projected screens are also used a great deal in digital presentations. Finally, presenters can hook up a digital presentation to a large monitor or several monitors.

DISPLAY OF THE PRESENTATION

As Putman observed, "video programs played from VHS tape on a 25-inch monitor may be satisfactory for a classroom, but it will look like the pits on a 10'x15' rear-projected screen" (1997, p. 52). Perhaps one of the greatest challenges for a digital presenter is understanding the major impact "display" has on the success of a presentation. A little detail, such as room lighting, can destroy digital presentations using a rear-projected screen. A projector's lamp brightness depends on screen size, audience, and room lighting. These all affect the image brightness needs of the digital presentation. The darker the room, the brighter the images appear on the screen. This makes it difficult for viewers to take notes and almost impossible for them to see the presenter. If the presenter's role is pivotal, then monitors are a better choice for display. The presenter ought to pick monitors that can reproduce a wide gray scale value, so that more colors can be perceived. This helps increase the presentation's readability factor. The room lights can be on when using monitors, allowing viewers to take notes. The negative aspect of monitors is that they require a great deal of cables and can take some time to set up.

Another overlooked aspect of "display" is audio. All rooms have different acoustics, so a multiband graphic equalizer is a "must" (Putman, 1997, p. 56). This piece of equipment keeps the audio levels consistent. The speakers should be placed in a three-way cross-over pattern for big rooms. Different audio power is required for different room sizes. For instance, a room of 20 to 30 people requires 10 W to 20 W, while a room of 200 to 500 people requires 100 W to 200 W. It is also vital to place the audience away from the outside edges of a room, because the sound will be too loud there. If the audio levels are proper for people sitting in the middle, the levels will definitely be too loud for people in the outside areas.

The placement of the audience is also important from a viewing standpoint. The seating pattern ought to be in a 45-degree angle for front projection and a 30-degree angle from the centerline for rear projection (Putman, 1997, p. 58). This places the audience in the optimum viewing and hearing area.

THE ART OF DIGITAL PRESENTATIONS

As Carl J. Luna and Joseph McKenzie observed, "multimedia's real strengths as a teaching method may well lie in hard to measure benefits in attention span and retention levels" (1997, p. 79). Thus, it is imperative that the producers of this technology use valid design principles. The fields of art and graphic design have been around for many years and, although the tools might be different, the goals are the same—communicate information to people through a medium. The medium affects the message, but should it be the message? The bells and whistles of technology can dilute the message instead of convey it. The object of digital presentation is to deliver information quickly and concisely without losing the audience. Currently, technology aids in keeping the viewer's interest, but, as this technology becomes more common, it will be good design that holds that interest. Therefore, it is becoming even more important to understand and use this powerful tool to create well-designed presentations. As Stewart Brand, multimedia artist, observed, "once a new technology rolls over you, if you're not part of the steamroller, you're part of the road" (1988, p. 9).

Bibliography and Suggested Reading

Brand, S. (1988). *The Media Lab: Inventing the Future at* MIT. New York: Penguin Books.

Hofstetter, F.T. (1995). *Multimedia Literacy*. New York: McGraw-Hill.

Itten, J. (1970). *The Elements of Color: A Treatise on the Color System of Johannes Itten Based on His Book* The Art of Color. New York: Van Nostrand Reinhold.

Lauer, D. (1985). *Design Basics*. New York: Holt, Rinehart and Winston.

Lindstrom, R.L. (1994). *The Business Week Guide to Multimedia Presentations*. Berkeley, CA: McGraw-Hill.

Luna, C., and J. McKenzie. (1997). Testing Multimedia in the Community College Classroom. *Technical Horizons in Education Journal*, 24(7), 78–81.

McLuhan, M. (1964). *Understanding Media*. New York: McGraw-Hill.

Putman, P.H. (1997). Strutting Your Stuff. *Video Systems*, 23(3), 52–58.

Vaughan, T. (1994). *Multimedia: Making It Work*. Berkeley, CA: McGraw-Hill.

Vlcek, C., and R. Wiman. (1989). *Managing Media Services*. Englewood, CO: Libraries Unlimited.

Wong, W. (1987). *Principles of Color Design*. New York: Van Nostrand Reinhold.

8 Graphic Design for the WWW

Do you see what I see?

Graphic communication is a symphony in choices of what notes to leave in and what notes to leave out. There is no other medium more prone to excess than the Internet—specifically, the World Wide Web. The pedagogy of effective design development, "less is more," has often been ignored in the Web world. The nature of Web publishing lends itself to intemperance. Unlike print publishing, in which the more colors, pages, and images the more expensive the project becomes—in Web publishing, all these extras are essentially free. Once the initial storage space is secured and a connection to the Internet is established, the rest is veritably free. Due to this and probably because the Web has until relatively recently been a medium for programmers, not much attention has been given to effective communication design. The maturing of HTML reveals the story of how graphic communication is now an integral part of Web development.

In the beginning of Internet programming, HTML was fundamentally a set of structuring tags for textural information that could be linked. Later, because of the phenomenal growth of the Web, HTML coding was expanded to include rudimentary layout functions. Still, using HTML 2.0 in 1995, the first standardized specification of the code, to produce tabs and consistent grid structuring was a somewhat awkward prospect. As graphic designers and artists became a part of Web development, efforts were made to make the Web more amenable to printlike layout processes. The World Wide Web Consortium (W3C) continues to suggest extensions to the language, currently at HTML 4.0, including centering, text-wrap, tables, and style sheets. Each extension fine-tunes the layout control in Web pages and the W3C continues to update these specifications almost yearly.

Print Publishing versus Web Publishing

As mentioned, most of the cost involved in Web publishing is in the initial startup areas: renting space with an Internet service provider (ISP) and acquiring a computer station to develop an HTML-based set of pages or a Web site. On the contrary, print publishing is prohibitively expensive for small companies and practically untouchable for individuals, because the cost of printing is substantial, especially if color is involved. Larger companies have capacious printing budgets for advertising and promotion due to the complicated and expensive print workflow described in Chapter 4. The WWW differs immensely in its favor for the individual. All one needs to publish on the Web is computer access, a rented line to the Internet, and a simple text editor, most of which are bundled with the Web browser. As one can see, the cost to publish in print compared with publishing on the Web is dramatically different. Not surprisingly, many people, companies, and governments have taken advantage of the price advantage in the WWW environment.

Another difference between print and Web publishing concerns the audience or user. With print, a specific audience is targeted and most often exclusively courted by the producer of the media campaign. Consider billboards. Most billboards are raised along freeways or expressways. Marketers pinpoint the demographics of the drivers most often on that particular road, and the billboards are designed to exploit those demographics. Look closely at the changes in products and services advertised on billboards mainly due to the makeup of the neighborhoods surrounding them. A Web site, on the other hand, may have a target audience in mind, but the "traffic" to the site is potentially the entire world. It would be impossible for a print publication to reach a worldwide audience without being exceedingly expensive to produce. In other words, print publishing is limited to its physical makeup and unchangeable data structure, while Web publishing is fluid and has no such limitations other than the equipment itself. However, a magazine can be read anywhere and doesn't require expensive hardware to access. Moreover, Web pages can be corrected, updated, or completely redesigned in an instant, compared with the time and money it requires to revise a print project.

A primary advantage of Web publishing over print publishing is in its use of hypermedia. Hypermedia is not merely the linking of text to other text (hypertext) but also the network of media elements connected by links on the WWW. Multimedia has found a comfortable home on the Web. The ability to follow linked multimedia elements in an associative process transforms the Web into an entirely new visual communication experience. However, hypermedia has been around for public consumption since the development of Apple Computer Corporation's HyperCard™ in 1987, which allowed users to handle hypertext functions in a proprietary format. A browser-type program allowed users to create hypertext documents as stand-alone items. The weakness of this kind of system lay in the necessity for a particular platform to use it, as opposed to the platform-independent environment available on the WWW. Browsers on the Web are integrated programs that translate hypermedia and display it in the context of an HTML-coded document. Basically, print can be categorized as a linear medium where users page through or read information in a sequential fashion. Technically, the WWW is also a sequential medium; however, because the speed at which information (through links) is accessed makes the linear process transparent, the Web has been accepted as a nonlinear environment. Therefore, we can additionally surmise that Web publishing differs from print publishing in its immediacy. Users can view and respond to Web sites instantly through forms and e-mail. The illusion of immediate information at the end of every link creates an impression of both urgency and interconnectedness in a virtual space.

The advantages of Web publishing are tempered by a key difference between the way Web pages and print pages are viewed. With print, a designer can control how the layout is presented. In a printed national magazine, the text, graphics, illustrations, and photographs are identical throughout the "run" of the magazine. Each subscriber sees the exact same colors, layout, and text. On the Web, variables in page viewing are significant. Pages viewed on different platforms (Mac, Windows, Unix, and so on), monitors, and browsers appear differently, according to developer considerations, color, space, and user preferences. To remedy some of these discrepancies, most professional Web designers use tables to anchor graphic elements and recommend designing for the lowest common denominator: a 13-inch laptop screen (640 x 480 dpi), Web safe color palettes (hexadecimal equivalents to RGB color), and three main text sizes (headline, subheads, and body copy). Since the Macintosh platform tends to darken colors and shrink text sizes, it is wise to design text slightly larger than it is to be viewed and the colors perceptively lighter. See Color Plate 8-9 in color insert. Another consideration is in image creation. In print publishing, large, highly detailed, multilayered photographs and images are common, however, on the Web, these massive files are certain to prompt the user to click away because of the extensive download times. Web site viewers commonly wait only 7 to 10 seconds for links or download information before clicking away. It is mandatory for the Web designer to engage a user in this space of time or risk losing that viewer for good. Studies have shown that regular Web users are not any more impatient than other audiences, but they have been conditioned to navigate in a certain manner at a certain rate of speed. In comparison, in the Western Hemisphere, we are trained to read from left to right; Web users are taught to respond to specific visual triggers and expect reasonable wait times for accessing information on the Internet. These visual cues can be consciously employed by Web-savvy designers to predict a common user's reactions.

The many file formats available for graphics and photographs in print were explained in Chapter 4. Unfortunately, static images on the Web are limited to two formats: GIF and JPEG. As discussed in Chapter 6, other dynamic media, such as video and animation, can be accessed through a helper application or a third-party plugin.

We have been discussing the differences between Web publishing and print publishing, but there are also many similarities. Some of the similarities between print and Web are text composition basics, emphasis considerations, color coding, and image impact. As flexible as hypermedia may seem on the Web, basic graphic design principles, the Gestalt theories of visual perception, are still necessary to create an intended design. Viewers require "a point of entry" other than the physical URL (Uniform Resource Locator), or address, that got them there, to comprehend fully the focus of the content. A hierarchy of information helps the audience read and view the main and subsequent points of the content in a logical order. Color coding the headlines, subheads, body copy, and hyperlinks by using consistent hues is a great way to let users recognize the relationships among chunks of information in the site. For example, the designer can "train" the viewer to interpret red as the color for all visited links throughout the site. Unfortunately, many of the consistency conventions that Web designers use in the Web site can be overridden by the user with the preference options available in most popular browser applications. Still, it is good practice to use a consistent layout, color scheme, and text face to endeavor to achieve unity within the site. Whether one is called a Web designer/producer, a Web developer, a content creator, a Web master, an information architect, or a creative professional, it is necessary to be versed in the practice of effective visual communication to maintain high-quality design solutions.

PRELIMINARY QUESTIONS

Most people believe that the Internet and the World Wide Web are synonymous, when in reality one is part of the other. The WWW is simply one service offered by the Internet system. The Internet provides the following services: e-mail; newsgroups (a sort of online bulletin board); mailing lists, also called listservs (newsgroups in which messages are sent to e-mail addresses, rather than posted to a page on the Internet); and the World Wide Web (millions of hypertext pages). See Fig. 8-1. The WWW is unique because it allows users to view text color, graphics, sounds, animation, and video interactively. This hypermedia capability has made the Web enormously exciting and popular.

Now that we have examined the similarities in and differences between the print and Web publishing arenas, let's look at the developmental stages required in creating a Web presence. Begin by asking these five questions:

1. Will this product or service be enhanced by a Web presence?
2. What form should the Web site take?
3. Who is the intended audience?
4. What kind of information do I want to provide in this site?
5. Who will create the site?

The first of these questions refers to the need to decide whether or not the WWW environment is appropriate for the product or service. For example, it may not be advantageous for a small local restaurant to have a Web site, but a national chain may benefit greatly from the worldwide exposure the Web provides. If one's appearance on the WWW is for personal reasons, then it is probably not mandatory to have a Web site. Interestingly, it has been estimated that 85 percent of the sites online today are for making money, not for ego purposes alone, yet actually making money online has been very difficult for most Web companies. However, the impression remains that, if one has a product or service to offer, the WWW is the place to sell it.

The second question to ask before developing a Web site is about what form the Web presence should model, as a Web site or a Web portal. A site is a collection of related, hyperlinked pages usually concerning one subject, while a Web portal is a

FIGURE 8-1

These are the services offered through a connection to the Internet

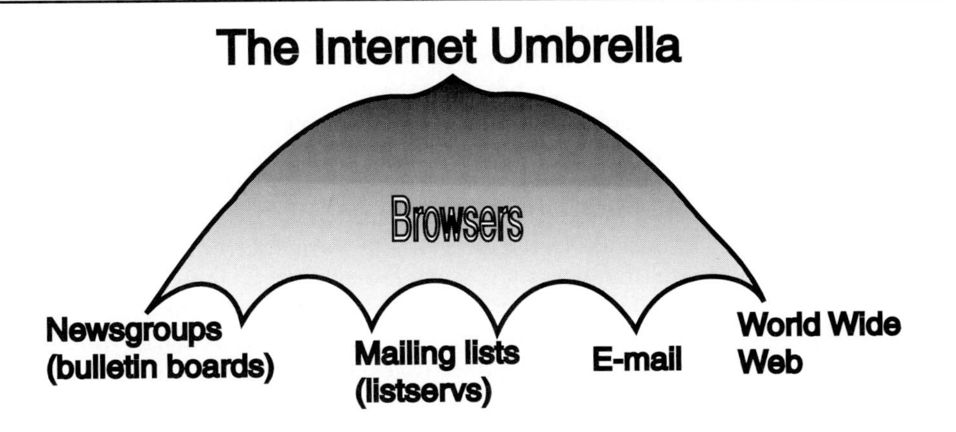

collection of many applications. The advantage in a portal is that server-side (at the location of the Internet service provider) programs can run within a single common interface. Some universities have recently started to offer portal-type services by allowing students to access their academic records, their schedules, and various forms online. Web sites can offer interactivity involving simple applications for accessing information in a database or search functions in the site but do not usually allow for running complicated applications. Regardless of whether one is designing a Web site or a Web portal, identical design considerations should be employed.

The third question to consider before building a Web presence helps one identify the target audience for the site. It is true that, when information is published on the Web, potentially the whole world has access to it. Nevertheless, a good Web producer should define an audience for which the Web site is being targeted in much the same way print publishers define their demographic constituencies. Recently, it has been approximated that 66 percent of all Internet users are over twenty-five years of age and those in the twenty-five- to thirty-four-year-old age range spend the most time online per month. These are important statistics for Web producers, because they begin to outline an audience profile. Obviously, it would be vital to research the typical users of the product or service itself to disclose a well-rounded picture of one's targeted audience.

Once the audience has been researched, the fourth question should be asked: what kind of information should be provided in the site? This question relates to the services or products being sold. There is an advantage for sites that provide a logical hierarchy of informational elements. If one is selling housewares, a varying visual emphasis and relational connections concerning the most important products to the less vital products helps the viewer make buying decisions. Shopping at online stores is often compared to the physical process of window-shopping. Consumers expect not only a presentation of the products but also a pleasurable experience.

The first four questions lead us to the act of creating a Web presence; we now need to ask who will actually create the site. Many people attempt to develop a Web site on their own. Although this may seem frugal at the outset, it is comparable to a homeowner attempting to fix his/her own plumbing. The homeowner may be familiar with using a sink or shower, but does the same knowledge guarantee wisdom about the inner workings of the pipes? Probably not, yet many Web users have developed their own sites with varying degrees of success. Some do-it-yourselfers come up with fairly competent sites, but they lack the sophistication and professional level of intended design needed to make the site both aesthetically pleasing and comfortably usable. Likewise, the homeowner may be able to tape a pipe temporarily to keep it from leaking, but eventually the pipe may burst. Consequently, it is wise to get a professional plumber to do the job. Similarly, just because one is familiar with using the Web doesn't guarantee competence in the effective development of Web sites. It is advisable to employ the skills of a professional Web consultant in designing a competent Web site. Web sites can be a wonderful way to advertise companies, promote products, and sell services, as long as they are well designed and carefully structured, especially in the area of site navigation. Because amateurs have developed many existing Web sites, a whole new cottage industry in site redesign has emerged. The novelty of having a Web presence, any Web presence, has waned, and taking its place is a consciousness about successful Web design communication. Web site owners are turning to media arts professional to design or, in some cases, redesign their Web presence.

Site Maps and Storyboards before the Code

Before any HTML code is written a Web producer must resolve the underlying purpose for the site. In the Web world, content is king. As amazing as interesting dynamic elements and interactive functions can be, the success or failure of a Web campaign is in the strength of its perceived content. Very often, because of the multimedia possibilities inherent in a Web experience, the main points of the site are lost in the "eye candy." A regular diet of sweets can make a body unhealthy, just as a Web site stuffed with irrelevant multimedia elements can render the site worthless. Audiences will consume the eye candy and come away with absolutely no knowledge of what the site is about. It is extremely important for a Web producer to establish the content priorities before any structuring code is developed. Outlining content, in an order that ranges from the most important to the least important information, will provide points of emphasis and levels of importance. Unsuccessfully navigating through the pages of a site for an elusive central point is all too often the Achilles' heel of an otherwise aesthetically pleasing Web site.

After the content of the site has been prioritized into logical categories, the actual building blocks of the pages should be organized into a permanent file order. In a Web project, it is necessary to organize the documents, images, animations, and text in one "root" folder. This is a mandatory step in the process because, when the site is uploaded to the server, the pages can then be found by a browser and displayed on the user's machine. Due to the nature of hypermedia, the linking functions for both the locations of elements on the page and what they may link to are interdependent. In other words, if one moves or changes the name of any file, the entire site may be affected because of the linking characteristics in HTML coding. When one encounters broken image icons or dead hyperlinks (which go nowhere), the culprit is usually a changed file name or a moved folder. Creating a root folder and saving all the text and graphic elements in the entire site before beginning the structural code will help avoid some of these problems. Site management is a built-in function for most Web-editing software used today and simplifies moving, renaming, and adding files or elements to a site. There are some naming conventions required to display pages on the WWW. The following is a list of naming rules to ensure one's files are published correctly:

- Stick to lowercase letters.
- Do not use apostrophes, colons, semicolons, bullets, slashes, or other special characters.
- Never use a space in any file name. An underscore can be used in the place of a space for clarity.
- All pages require a suffix or an extension (such as .html). Graphics should end with either .gif or .jpg, which refers to the file formats.
- Keep the file names short and name files for identification with pertinent meanings. For example, myface.jpg tells what the file contains much more easily than pict1.jpg.
- It is not unusual to have sites that contain hundreds of pages and thousands of elements, so keeping files organized with pertinent names and logical category folders is fundamentally important. Note that some ISPs expect files to be structured in a "flat" hierarchy, in which there is only one root folder and all

the files; graphics and dynamic content are located in it, with no subfolders inside the root folder. Others allow for subfolders, in which files, graphics, and dynamic elements can be organized into logical groups.

Once the root folder is complete with all text, images, and dynamic media intact, the process of laying out the pages into an easily navigable format must be tackled. Most Web designers create a visual diagram of the entire site, with all the dependent elements noted in the map. A site tree is one way of organizing this type of information. In a site tree, the pages are mapped in the way they will be linked together on the Web. The tree starts with the trunk, the homepage or index page, and proceeds to branch out to the rest of the pages in the site. All the linked elements, both pages and media, are indicated in text format for each page in the site. See Fig. 8-2. If the site tree is shallow and wide, it may mean that the information is unfocused and that some data should be organized into more specific categories to indicate category priority. Conversely, if the site tree is narrow and long, it may indicate that the information has been nested too deeply, and a user may tire of drilling down through too many pages to get to the desired information.

A flow chart indicating the way users will navigate the site is also desirable. This type of navigation map allows the designer to preview how a user might travel the site. Some designers recommend prototyping a site before publishing, much the same way manufacturers would engage in a test market prototype strategy for a new product line. A prototype is a clickable model, with placeholders for text and other media, that simulates the look and feel of the finished Web site. The prototype can be used as a tool to test the viability of the site links and navigation conventions.

A site tree, along with a navigation flow chart, helps define the emphasis of content and how users move through the site. A site storyboard is also necessary to judge the overall appearance and thematic integrity of the information. As mentioned in Chapter 6, a storyboard is a visual outline of the graphic elements in a dynamic visual communication project. A Web site storyboard differs only slightly from the trad-

FIGURE 8-2

A simple tree diagram for a venture capital company Web site

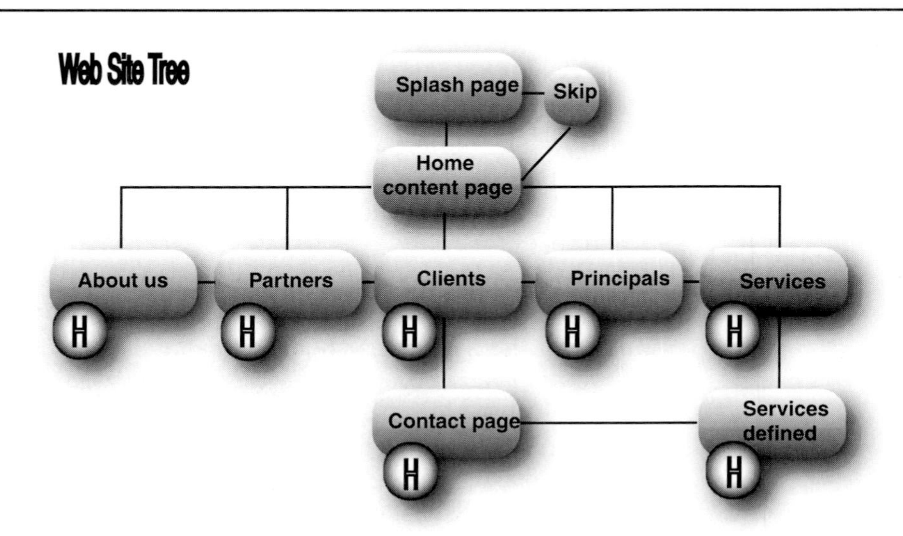

itional boards described in Chapter 6. A Web site may encompass three pages, as in a printed brochure, or it may bloat to 10,000 pages, unlike any typical print publication. This range in size is one of the aspects of Web design that is totally different from the kinds of projects encountered by traditional print and media producers. With the addition of interactive and rich media content (streaming video, audio, and animation), the Web medium changes the rules. A Web storyboard must include navigational linking elements as well as rich content considerations. For organizational purposes, these storyboards should branch out to indicate embedded media within each page. A panel in a Web storyboard might include the layout, text, and static images in the parent frame, with extending lines and related child panels to reveal rich media elements. Instead of following the linear pattern of a traditional storyboard, Web boards look more like an electrician's blueprint (see Fig. 8-3). The essence of a Web site is in its seemingly nonlinear menu of multimedia elements, with the capability of accessing information in seconds. The Web storyboard requires a designer/producer to visualize multiple graphic and media elements in an organic pattern, rather than a mechanical pattern in an attempt to direct that path for a viewer. Armed with this type of board, the producer is free to create unique visual experiences without becoming mired in the potentially endless possibilities.

Equal to the importance of a Web board is the producer's interpretation of what the site owner expects. A significant number of projects encountered by a Web consultant today involves the redesign of existing Web sites. The WWW has been around enough time for many nonprofessional designers to develop and publish Web sites on their own. It is not unusual for Web producers to face the challenge of revising "my cousin" efforts in developing a Web site for a client rather than the more enviable

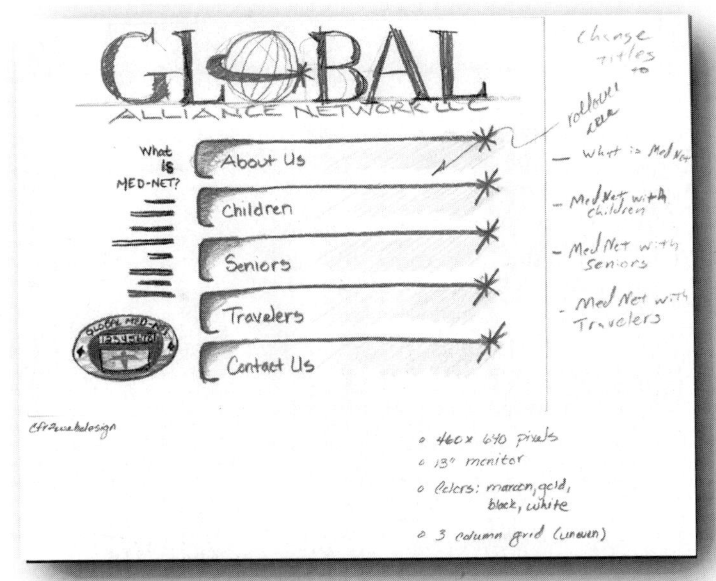

FIGURE 8-3
This is the homepage panel in a Web site storyboard for a medical data company called Global Alliance Network

position of starting from scratch. In either case, the preliminary preparations in the form of a site tree, a navigation flow chart, and a Web storyboard will ensure a level of sophistication in organization, layout and aesthetics that will far exceed any amateur's efforts. One should be careful not to underestimate the complexity of seemingly simple Web projects.

DESIGN AND LAYOUT CONSISTENCIES

After the HTML code, there are three visual elements in the anatomy of a Web project: the text system, the color system, and the page architecture system. All of these schemes demand consistency in that, once established, they must be followed to evoke a feeling of unity and relationship. When establishing a text scheme for a site, the producer should review the theme, or main emphasis, of the site. For example, if the project concerns a minor league baseball team, then the site should probably focus on the team's name, location, and game schedule. The text system could stem from the combination mark (logo and symbol) for the team and the players' professional activities. It is advisable to use type without serifs (sans serif fonts) for reading text, because the serifs, or the little strokes that project off the main strokes of letters at the bottom or top, tend to disappear in smaller faces, depending on the color combinations and screen resolution. In addition, serif typefaces usually have thick-to-thin strokes within the body of their letters, a throwback to hand lettering with ink pens, which, if extreme, break up on a computer monitor. It is possible for high-resolution monitors to display wonderfully crisp text faces in either serif or sans serif face, but therein lies the challenge of designing for the WWW—the volatility of the media. Users can define overriding preferences for colors and text at their browser level and the differences in color spaces for monitors, in computer platforms which can create uncontrollable variables at the designer's level. Some solutions to combat these variables have been invented, which we will discuss further in Chapter 9. At this point, we will examine some successful text typeface choices that can be implemented in most text composition instances. For large amounts of text, body copy, headlines, captions, and sidebars, Helvetica, Arial, Verdana, and Georgia are solid, easily readable choices. For serif options, New York, Century Schoolbook, and Courier can be displayed and read satisfactorily. Obviously, specialty fonts that are included in logos and lettermarks must be evaluated individually to determine legibility on a computer screen. It may be necessary to substitute more readable fonts from the printed version of the site's logo, rather than just digitizing the logo and slapping it on the Web page. As mentioned, some attention should be paid to the main focus of a site when choosing the text system for the publication. Our minor league baseball team could use dynamic or asymmetrically based text typefaces, rather than classic, symmetrical fonts to convey a feeling of action and movement. Simply put, use typefaces that reflect the nature of the site's emphasis and do not ignore the power of ambiance. In the final analysis, one of the most overlooked yet simplest rules to follow for successful type composition in a Web site is to limit type fonts. Include no more than two separate families (for example—Helvetica and New York), with no more than three variations (for example—regular, bold, and italic).

Establishing the color scheme and the page architecture in a site is comparable to developing the text system. Follow and limit: follow the theme of the site by choosing colors that reflect the psychological cues reminiscent of the subject and limit the palette so as not to bombard the viewer with unrelated color merely for the sake of color itself. Adhere to an underlying grid structure throughout the Web site and use sparingly elements that "break the grid," or layout of the screen broken into horizontal and vertical lines. For example, the minor league baseball team may want to limit their main color choices to their team hues—perhaps teal, orange, and magenta. This combination may sound garish, but remember that value and intensity also play an important part in color harmony. A high-value (light), low-intensity (less saturated) orange background combined with a consistent text system in a lower-value, high-intensity magenta will allow for easy readability and clear contrast. Add a medium-value, medium-intensity teal for graphic accents and a tertiary color scheme has been established. Colors in important photographic images can be the basis from which a color scheme is derived. Ultimately, the judgment of whether the colors work together to create a pleasing effect is subjective. One person's palette of beauty is another person's epitome of ugliness. Still, standard schemes based on the color harmony relationships discussed in Chapter 2 can provide a successful starting point for a Web producer.

The third visual element of a Web project, page architecture, is the organization of information into categories. The main pages are divided into sections that accommodate these categories and graphic elements. The grid unfolds from the content, not the other way around. These main pages are at the top of the hierarchy in the site, much as chapter pages are in a printed book. The subcategories for each main page should have a separate but related grid structure established from the informational and graphical components contained in these particular pages. This consistent page architecture yields a relational ingredient in the menu of pages contained in the site. The producer can then break the grid to create emphasis. Unfortunately, overusing this technique creates exactly the opposite effect; too much breaking of the grid structure destroys the consistency and therefore the organization of the elements in the site and confuses the emphasis of the information; however, used sparingly, these contrasting elements create interest and excitement.

The pages that make up a Web site are not all created equal. The first page in the site is the vital entryway. Even more complex than the design of a book cover, the homepage (index page) provides information about the subject as a whole and is the portal for all the relational subjects in the site. Some homepage necessities include the following: the site owner's identity, a navigation system, search capabilities, the site's contents, and the audience hook. It may seem obvious, but the individual, product, or company one is presenting must be identified on the homepage. Too many site developers are caught up in the technology of the medium and omit the main purpose of the site—to identify who, what, and where. Next to identification, it is mandatory to have a navigation system for the user to follow on the first page of the site. A table of contents in link format combined with a search field capability in a large site is elemental. Navigation bars are one way of satisfying this requirement. The ability to "hook" the audience, or keep the viewer interested in the site, is a much debated subject among visual communication professionals. Some believe that using dynamic content in the form of a splash page will intrigue the audience and lead the viewer into the rest of the site. Others contend that the long waits for this type of medium to download or the "extras" needed in third-party plugins for the browser to display the splash page animation or video null any initial interest or excitement the page may engender. In any case, it is wise to be aware of the site's target audience and attempt to match the technology typically available to the intended users.

HTML Coding versus Web Editors

It is not within the scope of this text to provide a complete explanation of developing HTML, but a brief description of how it works is essential to a Web producer. Since HTML is a set of tags that describe how a page is structured and not how content is displayed, it has been difficult for traditional graphic designers to embrace the medium. They are accustomed to creating unchanging layouts with minute specifications concerning text, graphics, and images. The Web is a fluid medium, with variables that cannot be permanently defined. With this in mind, most sites for the Web were developed in the 60's and 70's by hand coding in HTML, resigned to the limitations of the code. These sites tended to be boring and extremely volatile, because they were mainly text without variation, except in limited sizes and colors, and the pages looked different on each user's machine. With the development of additions to HTML and new mutations in coding, such as DHTML (dynamic HTML) and XML (Extensible Markup Language), graphic designers are able to match the flexibility and control of graphic design for print. DHTML expanded classic HTML at the 4.0 level to include layers, absolute positioning, and style sheets. Later, XML was created as a more flexible coding language related to HTML that includes tags and functions that can be defined by the publisher and displayed through a set of style sheets linked to the document. Although hand coding is beneficial in the designing of Web sites, it is no longer entirely necessary. Web editors, including Macromedia DreamWeaver™, Adobe Go Live™, and Microsoft FrontPage™ enable designers to drag and drop text, graphic and image elements in a layout without any HTML coding. These editors permit producers to include dynamic elements in the form of animations, rollover events, and interactivity without the need for extensive programming. It should be noted that, although these editors include scripted behaviors in the form of drag and drop components or templates, it is still advantageous for the producer to know the programming and scripting that creates these actions. The beauty of most Web editors is that the drag and drop interface favored by most graphic designers is complemented by an HTML editing component preferred by most programmers. For example, in DreamWeaver, as the drag and drop interface is being engaged by the graphic designer, a separate window with the corresponding HTML tags can be displayed for the hand-coding aspect of the page. The avenue a Web designer takes to structure a site is the first of many directions a producer of this media must attend to. Another main consideration in the map of a Web site is the production of graphics and images.

Graphics and Images

There are unique production requirements when developing graphics and images for the WWW. Notwithstanding other considerations, the size of the image should be the number one concern for a Web producer. Graphics and photographs for print media have the luxury of unlimited size constraints due to the improvements in removable storage components. Removable disks can be delivered physically to the media service providers, and the subsequent image files can be digitally stripped into prepared layouts. In contrast, Web images are stored on the Internet provider's computer, which then serves the file, in the correct format through the WWW, to the user's machine requesting the file containing those images. This digital journey ultimately requires files with images to be compact and easily transferred from computer to computer, avoiding prohibitive download times. Therefore, image files must be compressed yet still retain clarity and definition.

Many software manufacturers in applications such as Adobe ImageReady™ have tackled the task of preparing images and graphics for the Web. These packages allow producers to restrict color to Web safe palettes that include the 216 colors common to the main computer platforms in use today. In addition, indexed color palettes with optimized diffusions that simulate as many colors as possible are available functions in this software genre. ImageReady and other similar packages compress or optimize images and convert them to one of two image file formats the Web can display: GIF or JPEG. Images and graphics created in most popular paint and draw programs can be optimized for displaying on a Web page. There is also a special file format, PDF (portable document file), discussed in Chapter 4, which can also be displayed on the Web. PDFs have the unique ability to embed all layout, image, and font information in one file. The proper way to prepare a document as a PDF for Web publishing or for print publishing will be discussed more thoroughly in Chapter 9. Adobe Illustrator, Adobe PhotoShop, and Adobe Acrobat Distiller are three programs that can create and save files in the PDF format. Web site designers should be diligent in keeping graphics and images compact and within the Web's restricted color gamut while maintaining as much of the original image's richness and clarity as possible. See Fig. 8-4.

Original graphics and images can be obtained in a myriad of ways for both print and Web media. Images can be created from scratch, collaged with other images, purchased from a stock image manufacturer, or grabbed directly from an existing Web site. However, because of the ease with which images can be copied in the Web environment, some people seem to believe that copyright laws do not apply to the WWW. This is not the case. It is possible to be fined up to $150,000 each time copyrighted work is displayed publicly without permission, and this includes Web publication. As discussed in Chapters 5 and 6, the easiest way to avoid these penalties is to clear the

FIGURE 8-4

The same image, compared in GIF, JPEG, and PDF graphic file formats

File Formats

GIF format

JPEG format

PDF format

FIG08005 A.gif

FIG08005 B.jpg

FIG08005 C.pdf

rights, or obtain permission to use copyrighted pieces. How the work is used also makes a difference. Some work may be in public domain, or beyond the time it is under the protection of copyright laws, and can be used freely without permission from the owner. Still others may be in the realm of fair use. Fair use encompasses displaying images in nonprofit areas, such as in scholarship or education; work that is not particularly unique, or using portions of a larger work. In all cases, it is necessary to research an existing image or graphic to decide whether or not the owner requires permission. Generally, if one is using fine art or photographic works, it is necessary to obtain the rights to copy or distribute the work from the artist or audio/video producer.

DYNAMIC CONTENT

Most Web sites today enlist interactive elements beyond hypertext. Typical types of interactivity include fill-in forms, frames, plugins, Netscape's JavaScript (or Microsoft's ActiveX), and style sheets. A Web producer should be sensitive to the actual need as opposed to the mere ability to add interactive elements to a Web site. Fill-in forms may be the perfect solution for a site selling personalized jewelry, while being totally inappropriate for an informational site about an upcoming public event. "Mailto" forms provide instant user feedback by means of e-mail links. Using frames in a Web site permits the producer to group multiple pages in one screen. Each frame is a separate Web page, and the action of clicking on a link in one frame, as in the case of a table of contents frame, results in the display of a new page in another frame on the screen. Plugins are tiny helper applications that enable dynamic media. They can be inline, a part of the Web page itself, or external, prompting another window to be displayed in the browser. As mentioned in Chapters 5 and 6, QuickTime, Windows MediaPlayer, and RealPlayer are plugins that support dynamic media on the Web. Netscape's JavaScript and Microsoft's ActiveX scripting languages specialize in controlling Web browser processes, most often through event handlers. Event handlers are commands that trigger actions whenever a certain event occurs, such as the action of a mouse click triggering an animation to play. Finally, Cascading Style Sheets (CSS) are a standard method of creating formatting instructions and saving them for use with all of a site's documents, much like style sheets in a layout program. As one can see, the value of interactivity is twofold in a Web site. It can enhance a visitor's experience and can provide valuable information for the Web master.

One of the most popular kinds of dynamic content for a Web site is animation. Animated elements can be as simple as a button graphic changing color when it is clicked or as complicated as interactive splash page animation. Splash pages play animations, usually triggering the home page of the site. Graphic animations on the Web generally come in two forms: GIF animations or Flash animations. GIF animation is very similar to old-fashioned flipbook animation. It plays back highly compressed still images in a sequence that is compiled using a mathematical procedure. Still image creation programs that accommodate transparent layers are a viable option for creating the frames in GIF animation. Each frame is changed slightly from the preceding frame to produce the illusion of movement. See Fig. 8-5. The stack of files is saved sequentially and played back in a Web site without the need of a plugin. GIF animations can be added to Web pages in the same manner that still graphics are embedded in the HTML page. The simplicity with which this type of animation is

FIGURE 8-5

Six panels in a GIF animation for a Web design company

incorporated into Web pages often makes it the victim of overuse. However, the tasteful inclusion of simple GIF animation can make a Web site as serendipitous as an unplanned tulip in spring.

Flash animation is created with the use of vectors. The code for displaying each Flash frame is downloaded to the user's computer, rather than the entire bitmapped graphic for each frame, as in GIF animation. The image descriptions are continuously transferred in small packets of information to the user's machine from the server. The resulting impression is that of a smoothly moving animation. Since the animation code is not downloaded all at once, larger and more complicated animations are possible without the undesirable side effects of long waits and jerky movements. Macromedia uses the term *Shockwave* to refer to an entire group of players and authorware for its Web software, but Flash has become the most popular option with designers. Eventually, the Flash player plugin could be bundled with new-generation browsers, making it even more prevalent than it is today. Flash's interface consists of a drawing area, a palette of tools, and a frame-based time line. The learning curve for using Flash software is relatively low when compared with other vector-based animation suites, and this feature alone makes it very attractive to multiple-application-weary designers and producers.

Streaming content has further complicated the offerings of dynamic media in Web pages. Preloading is one way of creating a virtual streaming solution. Animations and videos are loaded to the user's machine and "Now-loading" screens are displayed to keep the viewer's attention while the full animation loads in the background. As discussed in Chapter 6, actual streaming incorporates a server running Web streaming software that packages the streaming information correctly and users possessing a large bandwidth connection to view the streamed files. The user's machine also needs a plugin player to view the streaming video or animation. Macromedia Flash and Adobe LiveMotion are two software packages that one can use to create rich media. In the end, a Web producer must be cognizant of the procedures needed to add rich media to a site, and the users must have the correct applications and plugins to display the animations.

BASIC WEB DESIGN RULES

As attractive as dynamic media can be, successful Web sites demonstrate a balance among information, interactivity, and entertainment. The following is a list of thirteen standard rules that most Web designers follow. Most professional Web consultants use these guidelines and probably have an extensive list of their own. These suggestions promote the beginnings of an effective Web design project while providing for individual flexibility in design tastes.

1. Realize that the average time a viewer stays with a site or link before clicking is ten seconds.
2. Avoid black backgrounds because, if the page is printed out, all the text will be invisible on a typical printout.
3. Use colors to enhance, not dominate, your Web page.
4. Stick to no more than ten links on the first page of your site, because most users do not go beyond this amount.
5. Be specific in the descriptive words included in the header of the site, because these are the words a search engine or directory applies when categorizing the site.
6. Attempt to display the most important information in the first 460-pixel x 640-pixel area, because this is the average screen size for the lowest common denominator monitors (laptops).
7. Avoid requiring the user to scroll sideways.
8. Top or left is the best place for navigation links, because this ensures that the viewer will see them in most screen sizes regardless of variables.
9. Give users viewing options by informing them before they click, especially how long it will take to download rich media.
10. Be selective about reciprocal links (hypertext to other related sites), because they normally lead viewers away from the main site.
11. Employ your consistencies to create a layout that is both visually appealing and easy to use.
12. Use contrast but limit it to places where emphasis is required.

13. Use restraint, which is fundamental: restraint in colors; text choices; hyperlinking; and, most important, dynamic media. If something doesn't enhance the purpose of the site, then don't use it.

It is fundamental for a Web consultant and producer to be well versed in these rules and the basic information in this chapter. Of all the millions of pages published on the Web, only a minute amount prompts repeat visits. Have you ever wondered why?

BIBLIOGRAPHY AND SUGGESTED READING

Baard, Mark. (2000). *Digital Video. Publish: Print, Internet and Cross-Media*. Topsfield, MA: Publish Media.

Baird, Bridget. (2001). 3D and Immersive Visualization Tools. *Syllabus*, 14(9), 23–26.

Bouton, Gary David, and Barbara Bouton. (1998). *Inside Adobe PhotoShop* 5.0. Indianapolis, IN: New Riders.

Butler, Susan P. (2000, August). Stay on the Right Side of Copyright. *MacWorld*, pp. 105–109.

Cohen, Sandee. (2000). *Web Invasion. Publish: Print, Internet and Cross-Media*. Topsfield, MA: Publish Media.

Cotler, Emily. (2000). *Designing for DotComs. Publish: Print, Internet and Cross-Media*. Topsfield, MA: Publish Media.

DiNucci, Darcy, with Maria Giudice and Lynne Stiles. (1998). *Elements of Web Design: The Designer's Guide to a New Medium*, Berkeley, CA: Peachpit Press.

Donnelly, Daniel. (2000). *Full Stream Ahead. Publish: Print, Internet and Cross-Media*. Topsfield, MA: Publish Media.

Evans, Tim. (1998). *Sams Teach* HTML 4: *Quick Steps for Fast Results*. Indianapolis, IN: Macmillan Computer Publishing.

Franklin, Derek, and Brooks Patton. (2000). *Flash 4: Creative Web Animation*. Berkeley, CA: Macromedia Press.

Gleason, Bernard W. (2001). uPortal: A Common Portal Reference Framework. *Syllabus*, 14(12), 14–15.

Gruber, John Edward, Jr. (2001). Graphic Design Is Not a Medium. *Syllabus*, 14(9), 27–28.

Langer, Maria. (2000). *Putting Your Small Business on the Web*. Berkeley, CA: Peachpit Press.

Laybourne, Kit. (1979). *The Animation Book: A Complete Guide to Animated Filmmaking—from FlipBooks to Sound Cartoons*. New York: Crown.

MaranGraphics. (Eds.). (1997). *Internet and World Wide Web Simplified*. Foster City, CA: IDG Books Worldwide.

Schmeiser, Lisa. (2001, May). Test Drive Your Web Site. *MacWorld*, pp. 44–50.

Snell, Ned. (1999). *Sams Teach Yourself to Create Web Pages in 24 Hours*, Indianapolis, IN: Macmillan Computer Publishing.

Williams, Robin, and John Tollett. (1998). *The Non-Designer's Web Book*. Berkeley, CA: Peachpit Press.

9 Default Design

Garbage in, Garbage out

Junk Communication

The absence of design is a hazardous kind of design. Not to design is to suffer design by default. We cannot afford to have graphics, products and architecture just happen.

Gregg Berryman

Gregg Berryman, a professor, an author, and a successful professional graphic designer for decades, is referring to the danger of inexperienced or, worse, inept producers creating and publishing visual communication without an understanding of what is being perceived by the viewing audience. Berryman further asserts, "Design is an urgent requirement, not a cosmetic addition" (1996, p. 2). The "default design" nemesis often causes problems concerning the "reading" of visual communication. In the best cases, designers are professionally trained in Gestalt theories of visual perception (see Chapter 3) and digital technology tools (see Chapters 5, 6, and 8) so that they can produce intentionally effective broadcast and published information for mass audiences. In the worse cases, the proliferation of digital tools that allows for laypeople to produce broadcast and published information at a fraction of the cost, time, and effort results in a plethora of junk communication. Although it has happened in the past, today it is more likely than ever that published material will be designed without much thought due to the prevalence of visual communication hardware and software tools. As it becomes easier and significantly cheaper to produce mass communication, many amateur designers are joining the ranks of published producers.

Junk communication can be harmless, as in the theoretical example of an eStore's homepage in which color schemes conflict and text composition is unreadable. This may be a losing proposition for the eStore's bottom line, but it is relatively harmless to the community as a whole. Unfortunately, designing by default can be dangerous. Suppose a hospital had signage that was developed by amateur designers because they happened to submit the most economical bid for the job (or, more likely, they were administrative assistants assigned the job). The

FIGURE 9-1

It's fairly obvious that this typeface is woefully ineffective for the purpose of the message

signage was difficult to read because there was little contrast in figure/ground color and the typeface chosen was a fancy script font. The script face was beautiful but impractical for medical signage. See Fig. 9-1. Most people who are looking at the signage in this kind of facility and trying to find their way around are not in a particularly keen state of mind. Remember, the people who were commissioned to complete the job did the best they were capable of doing. The ineffective signage was not intentionally confusing. After all, they were not professional designers. What did the hospital's administrators expect? In the end, the managing medical organization could be sued because there was the potential for patients dying before figuring out which way the emergency room was located. Although this scenario is ostensibly fictional, it is a tragic possibility when visual communication is not taken seriously.

This chapter is structured differently than the previous chapters. The process of evaluation should be a productive avenue for improving "mistakes" in media design communication—in particular, as media design pertains to media communication—rather than continuing further elaboration on how to use the digital tools. After rather lengthy discussions of the various multiple media and multimedia considerations in design technology, we come to a fork in the road. One path shows us how it is easier than ever to create, develop, and produce integrated media design and then publish or broadcast the results. The other path directs and teaches us about how to evaluate those beautifully functional, creative designs through a series of problems and their solutions. Assuming that you have mastered the accumulated knowledge from Chapters 1 through 8, this chapter will suggest solutions to media problems in content, color, the Web, graphics, video, typography, and finally craftsmanship. A media "mistake" will be explained and then a remedy will be suggested. These solutions, in the form of layout, graphics, video, and animation, can help creative media producers achieve effective and beautiful designs. Ultimately, the success of media content design depends on which path a creator travels.

COMMUNICATION TECHNOLOGY MISTAKES

Content

In today's production scene, creative content managers are in charge of managing staffs comprised of creative producers and technical experts while keeping up with the ever changing state of technology. Nevertheless, the goals are the same as in the past; these managers are concerned with developing a continual flow of visually compelling, clear, and functional material, be it printed, on the Web, or traditionally broadcast.

Problem: Inconsistency
Continuity in a transmedia campaign is vital. Often, it is easier to work in isolation from others in the creative process, rather than to collaborate your efforts. Component parts, including the print, Web, and broadcast media are usually created, developed, and implemented by different groups of people at various locations.

Solutions: Repetition and Communication
Repetition of colors, shapes, text, alignment, images, and sound are sure-fire ways to unify any creative production, whether it is a printed piece, broadcast video, or any other media production. A repeated color scheme or logo can develop a visual and therefore perceptual relationship between otherwise unrelated pieces of a media campaign. Not always obvious is the need for direct communication among directors, designers, artists, and production staff. In larger campaigns, it is even more essential to get people talking to each other throughout the process. Recently, print professionals have been sharing files in PDF format through electronic communication rather than face to face (ftf) during the production process. PDF files allow for editing on all sides of the communication stream, and, most important, these files are platform-independent. Adobe is making the creation of PDF files easier with each upgrade of Acrobat Distiller, and the Acrobat (PDF) Reader is free to all users. Also, to create professional printer files from third-party programs other than Adobe Acrobat when saving the file you want to convert, use the print function and "print to file" rather than sending it to the printer. This technique creates a PostScript file ready for use by a professional printer. As print professionals have discovered, communication through the World Wide Web or specific intranet portals is a cost-effective addition to ftf meetings and an efficient alternative for all creative professionals in a transmedia campaign. To summarize, follow these three rules for unified productions: repeat, communicate, and then communicate again.

Problem: Incompatible Platforms and File Formats
Standardization is a dirty word to most software and hardware CEOs, but ironically it would be the ultimate advantage for all users. Competing software and hardware companies find it in their best financial interests to avoid the standardization of platforms and file formats. Why make it easier for competitors to use their inventions? Further, many users of digital tools have not been formally or even informally trained in the utilization of equipment. They are expected to implement and present digital productions while trying to diagnose and fix breakdowns in the digital system

workflow. A telling example is an elementary school teacher using a word-processing program in conjunction with an imaging program for a creative student project. Most teachers are expected to implement rudimentary authoring programs into their curriculums. Certainly, this is not a problem when there is adequate technical support and the teachers have been sufficiently trained. More likely, printers stand idle because no one knows how to change the ink cartridges or images cannot be imported because they were developed on a Macintosh and the text program resides on a Windows-based system.

Solutions: Standards

Although standardization is a larger issue in the realm of digital environments, there are some simple rules that can be followed to remedy most incompatibility problems. Be aware of what type of program you are working in and how that software fits into the sequence of development tools. Rest assured that the same software manufacturer would support the component programs in their repertoire, but not a competitor's software. If using different software programs on different platforms—the Mac/MS Windows problem—then identify your file format and use only formats that are compatible with the software and platform you are using for your production. TIFF (.tif) files are recognized by most computer operating systems for images. Use JPEG (jpg) or GIF (.gif) files for electronic presentations (MS PowerPoint) or online work (WWW)—refer to Chapters 4, 5, and 6 in the sections on file formats for more information. Find an image-editing program that allows for the conversion of files from one format to another, such as Adobe PhotoShop, and use it to change file formats. When all else fails, attempt to cut and paste your images into the word-processing software you are using, because this process usually reconciles the differences in the file formats. Sometimes using a third intermediary program will do the trick. For instance, an obscure text file from an older program that imports into MS Word can be, in turn, imported into a layout program, such as QuarkXpress. Try to stick to reciprocal software between platforms—if it works in MS Word on a Windows platform, then it should be readable in MS Word on a Macintosh platform. Technical professionals should address problems with hardware, but frequently a brief perusal of the hardware manual's troubleshooting section provides adequate solutions.

Color

Color is amazingly effective. It is the first element audience members see when viewing a visual and usually the aspect they remember longest. Often, color is considered a frivolous concern even when it is the overall visual effect of a design that gets the first responses from an audience.

Problem: Arbitrary Color

Choosing color arbitrarily in the design process is a major mistake. Most successful media design professionals spend a significant amount of time and effort choosing color for their communications, and for good reason. Color layout and images are more affordable than ever, and making the mistake of using colors merely because you "like" them or because the colors are "what we always used before" is, once again, designing by default. Color should be designed by intention, like every other aspect of a media communication.

Solutions: Focused and Planned Color Selection

Be aware of basic color psychology and apply time-tested color schemes—refer to Chapter 3, in the sections on psychological color principles for an understanding of color perception and physical color properties for the various basic color schemes. Many studies and experts have proven that the most universally popular color is blue, especially "bondi blue," the shade created by Apple Computers. See Color Plate 9-10 in color insert. According to an online survey conducted by Colorcom, a color consulting Web site company, another favorite is metallic silver. Nevertheless, one must realize that color is intensely personal, and color tastes change with demographics, time, and trends, so diligence in researching hue choices for all visual communications is recommended. Make it a practice to choose colors that reflect the content of the communication, rather than your particular tastes.

Problem: Cultural Color Ignorance

It has been suggested that one of the reasons EuroDisney was originally unsuccessful was that the theme park signs used purple, which in predominantly Catholic Europe symbolizes death and crucifixion. Many visitors to the park described the signs as morbid and solemn. Obviously, this is not the impression Disney wanted to associate with its theme park. Reportedly, the CEO liked purple. The need to comprehend cultural color associations is mandatory in a global market. With the growing prevalence of the WWW, cultural sensitivities in a worldwide audience are essential.

Solutions: Color Associations

Becoming aware of the differences in cultural color associations is practical and rudimentary research for anyone designing media communication. In general, do not use the color of mourning for that culture. For example, black is the color of mourning in the United States, while white is the color of mourning in Japan.

The following is a general list of cultural color associations:

- *Africa*. The Turaeq people are nomadic tribes from Timbuktu who travel in the Sahara, and the only color they wear is indigo blue. However, the Bambara tribe in Mali, who are less insular than the Turaeq people, see dark purple, gold, and hunter green solids representing wealth, although they wear many colors. The Bambara wear bold, brilliant colors in celebrations. These two tribes represent the extremes of color associations in Africa and are a telling reminder that we cannot make assumptions about common color usage.
- *Asia and the Far East*. In Japan, the young, unmarried women wear soft pastels in pink and cream, while the older, married women wear black and beige. Black is the appropriate color for both weddings and funerals, even though recently Japanese brides have embraced the Western trend of wearing white. Red and white mean happiness and good luck. Purple and lavender symbolize the Japanese imperial dynasty. The Chinese are still more insular than the Japanese people, and accordingly the culture is more rigid in its color uses. White is reserved for mourning, and red is worn for weddings. Pinks, reds, and burgundies signify good health, good luck, and prosperity. Grays and navy blues represent the oppression of the Mao period and are not favored today. Colors in China are still associated with certain religious ceremonies and

superstitions. However, Western influences are appearing, especially in children's clothing.

- *England*. Traditional colors (blues, grays, and browns) in conservative tones still prevail in England. Most English people, with the exception of the young, who tend to break the mold in bold, bright colors, favor a muted and sedate palette.
- *France*. The French tend toward the lighter and brighter hues of the Impressionists, with the exception of Paris, where black is still a favored color.
- *Italy*. Italians favor bold Mediterranean colors, with a significant influence of the golds and purples associated with Catholicism in the Vatican.
- *Germany*. Germans tend toward earth tones in greens and browns, although, in major cities, bold primary colors are worn to attract attention.
- *Scandinavia*. Norway, Sweden, Denmark, and Finland do not have many hours of daylight due to their extreme northern latitudes; therefore, Scandinavians often use color to make up for the lack of light. Light blues, yellows, and bleached colors are prevalent in these countries.
- *United States*. Americans link certain color combinations to holiday rituals—black and orange for Halloween, red and green for Christmas. People in the colder climates in the nation tend toward darker tones, while those in the warmer climates favor lighter hues. Americans associate color in a consistently predictable manner, as discussed in Chapter 3.

Color on the Web tends to shift on different platforms (monitors, printers, and so on). Therefore, it is wise to use colors 10 to 20 percent lighter than you would in print graphics. Navy blues and darker tones often look black on a computer monitor, so add lighter and brighter colors as accents to avoid the heaviness that darker tones tend to imply.

Problem: Gradient Overdose—the "Because I Can" Syndrome

Just because it is easy and cheaper than ever to include a variety of colors in your designs doesn't mean that you always should. The use of gradients is a prime example. In the traditional print processes, the application of "rainbow" color banding is complex and expensive, but it can be as simple as two clicks of the mouse. The result seems to be a deluge of arbitrary color combinations in hideously overdone gradients. This is notorious in visuals developed with digital tools.

Solutions: The Color Wheel

Purchase a color wheel or color triangle and choose proven color combinations, especially if you are inexperienced in color design. Stick to color triads, three colors that are equal distance apart on a color wheel. Choose one main, or dominant, color and two accenting, or supporting, colors from this color scheme. See Color Plate 9-11 in color insert. Finally, do not always settle for template choices or default gradients, because they tend to be overused and garish. If possible, change the default color scheme to include a triad color scheme based on the content of your project, rather than the limitations of your defaults. No matter what media outlet the final visual will be published in, you need to limit your color palette for clarity of message.

Some useful color resource Web sites are

- Color Matters (*www.colormatters.com/entercolormatters.html*)
- International Color Consortium (*www.color.org*)
- Pantone (*www.pantone.com*)

World Wide Web

Among professional artists and designers, it is a "given" that, if you want to find an example of bad design, look to the WWW. The ability for anyone to publish on the Web has made it a breeding ground for ineffective, confusing, and downright ugly design. Regardless of this reality, it is possible to find some beautifully designed sites both in form and function. These sites avoid common design mistakes and embrace the Web as the unique medium it has become, rather than trying merely to copy print media of the past. The primary goal of a Web site is to get consumers to visit and stay at the site for a longer period of time. Surprisingly, WWW users spend about one-quarter of their time on a Web site viewing the organization's logo, so remember to include your logo.

Problem: Text Composition—Not Enough or Too Much
Many people who develop Web pages simply translate pages published in print form directly to the Web without considering the idiosyncrasies unique to this media environment. On the other hand, others overindulge and ignore the possibilities of "information overdose." The result is unending pages, unreadable text, nonprintable formats, and lack of content.

Solutions: Consistency
It doesn't cost per page on the Web, as it does in the print world. Use as many pages as necessary and break up your information into logical, easily understood chunks. Long scrolling pages are not only ineffective but also very irritating. The Web is a nonlinear medium; therefore, do not design pages that require text to be read in a strictly specific order to understand content. Each element of the layout should stand on its own and not rely heavily on the content contained in other elements on the screen. Conversely, do not underestimate the need for actual content and not just eye candy. Consumers visit a site for specific information and become dissatisfied very quickly if there is not a significant amount of content. Compose text so that the most important information lies in an average screen size, approximately 640 pixels x 480 pixels. Information can be outside this area, but it should be referenced in this prime screen space. See Fig. 9-2. Make available a print-friendly version of your page for the user, because many WWW users go to a site and immediately print the information, so that they can quickly surf to another site for more content. This immediacy is one of the major advantages of the Web. Aim for the middle by using text faces no smaller than 12 points and no larger than 16 points for "reading," or body, copy. Adjust the rest of your text (headlines, subheads, captions) to the body copy sizes. Last, check to ensure that your text color contrasts significantly from the background of the page. Reading on an electronic screen can be more tedious for the eyes than reading on a traditionally printed page. Above all, do not waste a user's time.

FIGURE 9-2

An area of 640 x 480 pixels at the upper left corner of the screen displays on a 17-inch monitor (diagonally) with a resolution of 72 ppi

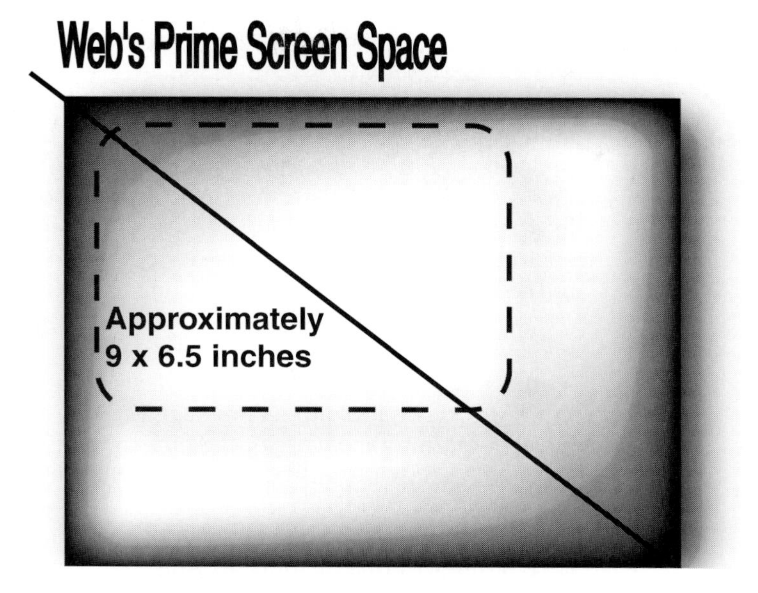

Problem: Graphics—Too Big, Too Long, and Too Dark

One of the major reasons users click away from Web sites is the time it takes to view high-resolution images because of inadequate bandwidth to handle these types of files. Huge images, in actual screen real estate, should also be used sparingly. Although contrasting size is a wonderful way to generate visual interest, many large, long downloading images is simply the wrong approach on the Web. Images that look wonderful in print do not always translate satisfactorily on the Web. Digitally scanned photographic images invariably shift in color and tone on the dark side.

Solutions: Clean and Concise

Optimize all graphics for Web viewing by first limiting color palettes to the 216 Web safe colors and second by reducing the images' physical size. Make the site attractive and persuasive by using graphics, but be very careful not to include nonessential images. Consider that some users "turn off" or cannot view graphics for various reasons and need to see an <alt> tag (alternative text if the graphic is not displayed). Refer to our discussion about the designing of successful graphics for the Web in Chapter 8. Always include a caption or descriptive text for every graphic link on your page. Generally, users will not click on graphic icons unless they have a text label. Photographic images need to be scanned expertly and edited so that they are easily read and relate to the content elements on the page.

Problem: Irritating and Unappreciated Animation

Animations can create exciting visual dynamics on a Web site but are also very time-consuming to produce. Too many times, Web site animations are irritating because they keep repeating or take too long to download. In some instances, users are virtually "trapped" by splash pages or entry animations.

Solutions: Use Animation When It Aids in Communication

There are two main options when producing animation for the WWW: GIF and Flash. Regardless of the file type, the main problem has always been file size. The faster you can get animations to download, the more likely your users will stick around to view them. Therefore, finding ways to optimize animation files is the key to solving animation problems. Flash animation is vector-based and can handle bitmap images. GIFs are exclusively bitmap where only the differences between frames are recorded. With GIFs, attempt to use transitions between frames that require the least amount of memory, such as wipes instead of dissolves. Use other transitions sparingly. Reduce the number of colors in the animation by applying the color reduction features of your GIF animation software. Reduce the frame rate and keep as much of the frame still as possible; animate only small things. Try using reveals—two images in the window and one progressively masks the other in turn—which is a great way to inject a dynamic aspect to your Web site. In Flash, use symbols as much as you can. Symbols are saved in the file once and can be used an infinite number of times and ways in the final animation. Consider saving your Flash file as a GIF animation, because everyone can view it without any kind of plugin. These are some of the ways to save file size in Web animation. Go on to explore all the size-saving opportunities in your software program. One simple concept you should keep in mind is to constantly evaluate whether or not animation is enhancing your message. If not, dump it. Create a storyboard, because it forces you to plan out how animation will work. Use movement for emphasis only and do not have two animations running at the same time unless one points to the other. Cover transitions with other movement, so that you do not lose the viewer's attention and try to set up an internal rhythm to the animation.

Problem: Linking Madness

A media producer was asked to evaluate a Web site for a minor league baseball team. The site contained approximately 50 pages, and it provided relatively viable content. Unfortunately, the producer was horrified by one of the main pages, because it contained no fewer than 200 hyperlinks! It was a chaotic mesh of link after link after link. This is madness, the producer thought. Who would be crazy enough to sift through this mess, let alone design it?

Solutions: Limit Your Links per Page

A user spends approximately ten seconds per page unless something catches his/her interest. Because of this impatience, having more than seven to ten links per page is not advisable. There are instances in which this rule can be broken, as in a frame-based page with a table of contents, but overall too many links are confusing for the user. An excess of choices makes the user uncertain about or, worse, apathetic to your message. The key to effective Web site linking is intentional, well-thought-out navigation, and this starts with the logical breakup of information. Take time to categorize your information carefully and to present it in attractive packets of interesting content. All "dead links," links that go nowhere because the reciprocal page is no longer in existence, should be eliminated by regularly checking your site specifically for them.

Problem: False Interactivity
False interactivity is usually not intentional. It occurs when graphics, text, and images look like interactive choices but, instead, are "static imposters." How many times have you clicked on a graphic only to be frustrated because it goes nowhere? It was never supposed to go anywhere. The Web developer thought it looked nice. Or consider instances in which a text field box turned out to be just another overly thick rule line. It was never supposed to be a type field. The Web developer thought it looked cool.

Solution: Do Not Mislead the User
The obvious solution for false interactivity is to be acutely conscious of how graphics are perceived. For instance, if you put a drop shadow or an inner embossment on a small graphic shape, it tends to read as a button to be clicked. If a line becomes so thick that it looks like a box, then it is a box. Also, the smaller the photographic image, the more it looks like a link. Even with the relative newness of the WWW, users have been conditioned to look for commonly accepted graphic conventions. Do not try to reinvent these conventions. Use this conditioning to your advantage by planning images carefully and becoming graphically aware.

Problem: Irritating Banner Ads and Pop-Up Windows
Many site designers use JavaScript to program extra browser windows to open automatically on a user's screen. They open when you enter the site, shortly thereafter, or as you leave the site. Along with banner ads, these uninvited windows are, at the least, annoying and, at the most, a reason to never visit that site again.

Solution: Do Not Use Pop-up Windows
Since most pop-ups are created in JavaScript, you can prevent these windows from making their irritating appearances by turning off your browser's JavaScript support. Web designers should never include them in Web sites, unless alienating users is their intention.

Image Capture and Audio/Video

Graphics

We have addressed graphics fairly extensively in the previous chapters, but there are some recurring problems with graphics in a production context.

Problem: Pixelated Scans Most image problems begin at the scanning stage. The weaker the scan, the more difficult it is to reconcile in the image-editing stages. Too little resolution and the scan is blurry and shows aliasing, or jagged edges. Too much resolution and the scan is much larger a file than it needs to be for its final output.

Solutions: Resolution Simply matching the pixels per inch to the scanner's sensors can solve the problem of calculating resolution. Choose a resolution equal to or less than the scanner's true optical resolution. Optical resolution is the smaller of the two numbers used in describing a scanner's dpi—600 x 1,200 dpi indicates an optical resolution of 600 dpi. The second number—in this case, 1,200 dpi—describes how far the stepper motor is able to move the scanning bar—1,200 steps per inch. Avoid the scanner interpolating (guessing) pixels. This pixel averaging process softens an image; consequently, the image loses clarity. The exceptions are line art and black-and-white line drawings. With these kinds of graphics, let the scanner interpolate 1,200 dpi, or you'll get pixelization on the imagesetter. Always scan in RGB and later convert to CMYK after image editing, because the scanner itself is an RGB device and needs to do less translation this way. In addition, RGB mode creates a better archival file.

Problem: Color Shifting and Calibration Problems Color shifting and calibration problems are common in the digital imaging production process. The different color spaces of scanners, monitors, printers, and digital cameras need to coincide to produce acceptable image quality.

Solution: Calibrate Your Equipment Getting scanners/cameras, monitors, and printers to agree when displaying color can be achieved by calibrating different devices to use the same ICC profile (International Color Consortium—*www.color.org*). An ICC profile is a description of how a device converts an image from the CIELAB (Lab) color, which is device-independent, into its own color space and back again. The Lab version of an image is retained throughout its journey in the production process. Profiles can be created for each device in the production chain—scanner/camera (input), monitor (display), and printer (output). These profiles can be created in-house by using a target of color swatches—IT8 is the most common target used today—or by using a third-party firm to generate output profiles from calibration files you have created on specific devices. Note that accurate profile creation requires a spectrophotometer color measuring tool and software to convert the data to a device profile. In the near future, color calibration processes will be invisible to the designer, because they will be embedded into software and standardized for each hardware device, but, once again, the concept of standardization pops up.

Problem: Digital Photography Image Quality In the prepress field, there is widespread acceptance of digital camera photographs, but professional studio photographers have been slow to accept this new tool. They argue that the prices are too steep for the high-end digital cameras that produce image quality already available with their film cameras.

Solution: Higher Resoultion Inevitably, as prices lower and megapixels increase, professional photographers will be forced by their clients to embrace digital cameras. The simplicity and immediacy of digital photography allow for clients to be in on the selection of images as they are being taken, rather than after a long wait, as in the traditional film-based proofing process. Lower-end digital cameras can be used by output venues that require less resolution, such as the WWW, newspapers, and television. Higher-priced digital cameras have steeper resolution requirements, as do magazines and brochures. In any case, digital photography will eventually usurp film photography as the medium of choice for commercial imaging.

Digital Audio/Video

Problem: Television's Limitations Because of television's technical standards, which adhere to mandates requiring color TV to transmit a signal that black-and-white sets can receive, television's video signals have been compromised. Many digital video producers are ignoring the limitations of TV and producing on-screen graphics and titles that not only look terrible on TV but are actually illegal because they create a signal that does not conform to FCC regulations. TV uses a scanning method that is interlaced, every other line, while computer screens paint the image's horizontal scan line sequentially from top to bottom. Consequently, thin lines disappear on a TV screen, and strange artifacts appear when translating computer graphics to television.

Solutions: Multiple Formats Transmedia production necessitates producing video that plays seamlessly in multiple media—television, movies, CDs, and the WWW. TV's interlace display has problems with fine horizontal lines and patterns; therefore, it is necessary to avoid lines that are only one pixel or an odd number of pixels thick. This alleviates disappearing lines, strange flickering, and random artifacts appearing on the TV screen. For the best text results, use a sturdy, bold, sans serif font, such as Helvetica medium or bold. If you must use a serif font, choose one that has less contrast between thick and thin strokes within the characters, such as Century Schoolbook. To avoid color problems, use only NTSC-legal colors and omit heavily saturated colors that are prone to "blooming," or spreading, into adjacent scan lines. Keep all important images and text within TVs "safe zone." See Fig. 9-3. Most video-editing programs provide a feature that shows TV's safe zone.

Problem: The Web's Download Nightmare Digital audio/video requires a great deal of storage space, rendering it fairly incompatible with the limited bandwidth of the WWW. Nevertheless, Web designers include digital audio/video in their Web pages, some requiring hours of downloading time. Once a user clicks on the video link, the only way to bail out of the download is to restart the computer system. Additionally, at this time, digital audio/video cannot be handled without first downloading third-party plugins.

Solutions: First, make sure the video is absolutely essential to the content of your site. Consider the possibility of delivering your audio/video by other means, such as

FIGURE 9-3

A diagram defining TV's "safe zone" (about 5 percent inside the outer edge of the screen)

a stand-alone CD. If the audio/video is deemed vital to the site, optimize the images and sound clips to the smallest file size possible, being careful to maintain acceptable clarity. Ensure that, if the user clicks on the audio/video, he/she is able to bail out of the download without having to leave your site. Include text informing the user about the video link and how long it will take to download on a variety of Internet connection speeds. Addressing the plugin requirement, digital audio/video producers must balance the pros and cons of the available players. Measure RealVideo's advantage as one of the most direct routes to a large Web audience with Apple QuickTime's delivery of a better moving image and smooth playback without the strange image freezes associated with streaming video. Consider using the Sorenson Video format, developed by Sorenson Media in Salt Lake City, because support for this low bit rate codec is built into Apple QuickTime version 3 and later. Or choose VBR Encoding, which is a two-pass analysis and compression process that creates movies with variable data rates. The most complex parts of an audio/video clip are assigned more bandwidth by the encoder, thereby creating faster downloads and smoother playback.

Text and More Problems

Typography

Problem: Conflicting Typefaces It seems that typographic design has been demoted to an afterthought. A disturbing amount of visual communication today demonstrates a woeful lack of attention to type design within a layout. Typefaces conflict with each other, rather than harmonize to promote content.

Solution: Variety is the Spice of Life Use contrasting or concordant type combinations. A contrasting type combination occurs when the typefaces used in the layout are limited and completely different from each other. The text composition is visually interesting, because the body copy and headline faces are completely different—for example, use an Old English font for the headlines and a modern Bodoni font for the body copy. Contrasting combinations have to be very different in structure, shape, and color. Be careful about choosing faces that are only slightly different—the faces must be distinctively different to be a successful contrasting text composition. Conversely, concordant type compositions use exactly the same face for all text in the layout. It is imperative that all the text be composed in the same typeface, but differences in size and variation are acceptable. Although this kind of text composition may be somewhat boring, it is always harmonious.

Problem: Text Insensitivity Text that is chosen arbitrarily independent of the message or mood of the content can be jarringly unattractive and annoyingly incongruent.

Solution: Be Text Sensitive The content of the message through thoughtful text selection and control of the color, weight, structure, and shape of the typefaces. We already know how color affects us psychologically, and it works the same way with text color. Additionally, the way each letter is built—its stroke weight, character structure, and overall shape—constructs an emotional impression on the viewer. Being sensitive to typefaces allows a designer to form a complementary collaboration between content and text, thereby strengthening the overall message.

Problem: Ignoring Negative Space Filling every possible space in a layout creates a tacky, unsophisticated, tedious mess of practically unreadable content. This is normally attributed to inexperienced designers or those wishing to cram as much information into the smallest amount of space for economical reasons.

Solution: Be Aware of Negative Space Do not ignore the power of negative space. Remember the old Buddhist saying that music is not in the notes but in the pauses between the notes. It is a huge misconception that the more information you can stuff into a layout the more effective the message. Ironically, the opposite is true. Less is actually more when it comes to type composition. Divide large galleys of type into digestible chunks of information and leave as much negative space as physically possible. On the Web, screen space is not an issue, so it is especially nonsensical to omit negative space in this medium. In all media, try to commit at least one-third of the layout to wonderfully effective background respites of negative space.

Problem: Missing Fonts Many times, inexperienced designers will use computer fonts that are incomplete in the context of the printing processes.

Solution: In explanation, there are essentially two types of font formats: True Type and PostScript. True Type fonts are missing the outline description necessary for printing the font to a laser printer or an imagesetter. This results in missing font errors and type that simply will not print out. Make sure that you use only PostScript fonts when generating designs that will be printed out in hardcopy form. Even if the output is electronic, it is wise to limit yourself to PostScript fonts, because both the screen description and the outline print description are embedded in this file type.

Craftsmanship

Problem: No Vices/Amatuers The proliferation of untrained producers and designers who are ignorant of basic design principles and possess no historic references are the main culprits in designing by default. Ignorance is not an excuse when potentially worldwide-published pieces are the result. Media are used in haphazard, unethical ways as a means to an unplanned end.

Solution: Training Experienced producers and designers must provide affordable training in both the theory and technology of media design communication. If you cannot take a class or at least attend a seminar covering design and content issues, than pick up a book. There are thousands of discipline-specific technology manuals and an equally large number of design theory-based texts. Look at the bibliographies of this book and pick out a couple of books to read. Enhance your skills by any means open to you.

Problem: Unknown End Use for Mediated Message Designs that are produced without the end in mind at the beginning inflate costs and dilute content.

Solution: Know Your Audience Develop a sense of the whole. When working on the parts of a design, whatever the medium, keep in mind the ultimate goal of the project.

It is tremendously easy to get caught up in the technical tools and digital toys associated with a media design campaign. Find a way to focus on each part of the process while maintaining a healthy feel for the final outcome.

Problem: Technology Failure Every digital tool has failed at one time or another. What happens when the technology is not available?

Solution: Prepare a Backup Plan Always have a nondigital backup. Use slides and photocopy handouts in the place of digital presentation or print out proofs. Communicate with your clients by any means possible and do not make excuses for the technology. Be ready to explain your ideas verbally if necessary. Finally, prepare for financially compensating disgruntled clients. The design of media communication is sometimes a merciless job, and new technology, by its very nature, is often unstable or unpredictable. Being prepared for all foreseeable instances is the best way to reconcile the variables.

Default design is a persistent problem in media communication today. With the convergence of media, traditionally trained professionals are being forced to embrace new technologies and paradigms, whether they are ready or not. At the same time, nonprofessionals are diving into the pool of design at an alarming rate. The primary goal should be to find aesthetically sound solutions using traditional and digital media technologies. In the final synthesis, one does not have to invent new design principles, just find new ways to combine existing ones.

Bibliography and Suggested Reading

Berryman, Gregg. (1996). *Designing Creative Portfolios*. Menlo Park, CA: Crisp.

Cooper, M., with A. Matthews. (2000). *ColorSmart: How to Use Color to Enhance Your Business and Personal Life*. New York: Simon & Schuster.

Dayton, L., and M. Gosney. (1995). *The Desktop Color Book*, New York: Henry Holt.

Flanders, V., and M. Willis. (1996). *Web Pages That Suck: Learn Good Design by Looking at Bad Design*. Alameda, CA: Sybex.

Groff, Vern. (1990). *The Power of Color in Design for Desktop Publishing*. Alameda, CA: Management Information Source.

Guidice, M., and A. Dennis. (2001). *Web Design Essentials*. (2nd ed.). Berkeley, CA: Peachpit Press.

Heller, S., and T. Fernandes. (1999). *Becoming a Graphic Designer*. Alameda, CA: John Wiley and Sons.

Itten, Johannes. (1996). *Design and Form*. New York: John Wiley and Sons.

Joinson, Simon. (1999). *The Digital Photography Handbook*. London: Duncan Petersen.

Kentie, Peter. (1991). *Web Design Tools and Techniques*. (2nd ed.).Berkeley, CA: Peachpit Press.

Langer, Maria. (2000). *Putting Your Small Business on the Web*. Berkeley, CA: Peachpit Press.

Marquand, Ed. (1986). *Graphic Design Presentations*. New York: Van Nostrand Reinhold.

Tuckett, Simon. (2001). Web Animation. *Step by Step Graphics*, 17(5), 109–113.

Williams, R., J. Tollet, and D. Rohr. (2002). *Web Design Workshop*. Berkeley, CA: Peachpit Press.

(2001). Color Handling and Upscale Scanning. *Step by Step Graphics*, 17(5), 20–23.

(2001). *Handbook of Pricing and Ethical Guidelines* (10th ed.). New York: Graphic Artists Guild.

10 New Technology

A need—a design—a selection

The thought-provoking comment often attributed to the creative people at Disney Studios maintains "you can teach an artist to be a technologist but you cannot teach a technologist to be an artist." Creative professionals who are rapidly learning to use digital tools with an eye for future possibilities are implementing these new technologies in cross-media communication. Technologists are realizing that creating the hardware, software, and networking is merely the beginning of effective digital communication. In the "Class of 2005 Mindset List" (an annual compilation of fifty items indicating the viewpoints and frame of reference of entering college students assembled by Beloit College in Beloit, Wisconsin), some interesting cultural facts about our future establishment are presented. Number twenty-nine (out of the fifty), states that the freshman class of 2001 have always used e-mail, and number forty-two jokes that, to these students, *beta* is a preview version of software, not a VCR format. Further, most students starting college in the fall of 2001 were born in 1983, the same year the PC and Macintosh systems were born. It is evident that the present population of college freshmen is, at the least, technologically savvy, if not masters of technology. Therefore, it may be safe to assume that simply having technology experiences is a given, and the challenge is couched in the effective uses of these experiences.

The guts of a media campaign lie in the successful design of user interfaces and communication portals. Rather than limit their clients' communication to specific media, designers and producers are expected to foresee publications in a variety of media outputs. Print media publishers call this mindset shift in design and development transmedia communication. See Fig. 10-1. Transmedia communication is about output convergence in that a communication solution to a problem is expected to be ported to many different outputs at the same time. No longer is it wise to design for one medium. Designers and producers must have an eye for the end product in both form and medium. With the concept of transmedia communication as a basis for all future designed solutions, new technology can be considered a process in itself as well as a tool. We can examine a media campaign's process by defining the following factors: a need, a design, and a technology selection.

FIGURE 10-1
In a transmedia model, media development overlaps

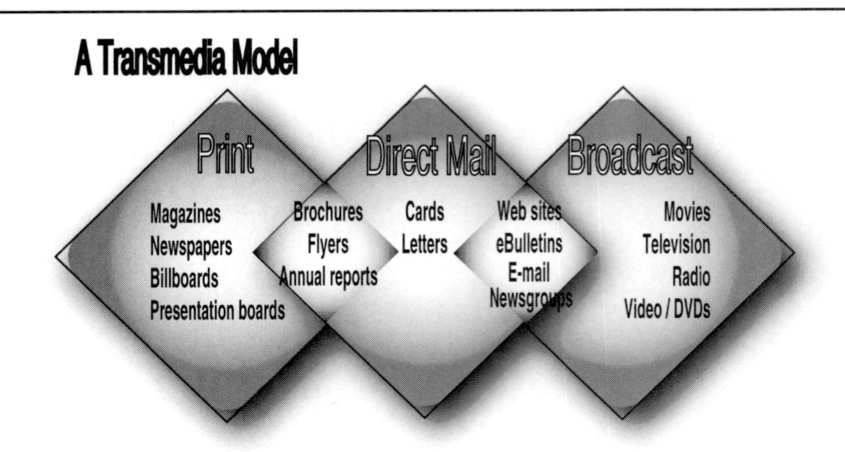

A NEED

The increasing demand for Web sites to accompany print media publications is a dynamic example of the surge in cross-media development. In the not so recent past, clients who desired a published presence usually tapped a marketing or an advertising firm to develop a cohesive print campaign. The creative company worked closely with the client to determine its target audience, product or service branding, and one or two media outlets that best served that client's needs. At the outset, the print designers considered various media for possible publication: magazines, newspapers, tabloids, direct mail, and billboards. Meanwhile, a separate broadcasting company developed a campaign for the same client, with considerations in the various outputs for moving media: TV commercials, movie trailers, animations, infomercials, and video productions. Today, the same client can approach one company, where all the various media outlet developers are housed. Often, the same creative professional who develops the print campaign also creates the animation and Web pages for the campaign. The filmmakers and videographers who develop the client's traditional TV broadcasts are most likely creating a digital video, which will be available on the client's published Web site. That digital video is saved in different formats so as to be available in analog videotapes through multimedia interactive CDs, which can be distributed to customers with the direct mail printed pieces. The definition of a client's "need" has changed from the question of which media it chooses to how many ways it can express its presence through media convergence.

Once the media design company has pinpointed the various integrated media choices, it is mandatory to have a collaborative plan of action. In the old days, for example, a graphic designer could work for a client in isolation, oblivious to the other peripheral media being created for that client. Certainly, the designer studied the client's previous printed media, but intimate knowledge of other media outlets was often irrelevant to a successful print campaign. In the climate of embracing transmedia communication campaigns, it is vital for every producer in the process to be completely aware of all aspects of the media design communication campaign. Each producer needs to complement every other producer's creations to ensure the cooperative use of both graphic resources and unified messages. As discussed in earlier chapters, sometimes one producer is responsible for all aspects of the media design from print to broadcast. The important factor is the design of the message which must be coordinated and focused on each medium. The coordination of these media designs begins with understanding the void that needs to be filled for the audience. A media producer fills this void by coordinating multiple tasks.

The tasks of the campaign are broken up into manageable micro projects, which are then assigned to various producers to complete. This ensures that designers within the campaign are working on complementary components, rather than on redundant elements. Let's say a new type of hybrid potato that cleans a person's teeth while being consumed has just been invented. The company that holds the patent for this "clean potato" wants to trumpet its virtues to the public in a marketing and advertising endeavor. The inventor chooses an advertising company. At this point, the information architects within the ad company begin the work of dividing up the components of the clean potato campaign. The patent holders expect that the clean potato will need a massive media blitz to establish it as a desirable product by the public. The producer conducts a needs analysis, which usually takes the form of a survey. This survey is given out to the prospective audience to help the producer develop a design with maximum impact. Focus groups may also be used. This analysis presents the product to a small group of people that represents the demographics of the larger audience and then analyzes their reactions. A media design will be based on the information gleaned from the needs assessment and research on the target audience. See Fig. 10-2.

A DESIGN

Biologically altered food tends to have a negative connotation among the grocery-buying public. This perception has to be changed through successful media communication directed to the consumers. The challenge for media developers is to use the positive aspects of the product to overcome the negative connotation of an artificially altered vegetable. It is essential that the producers of such a campaign take

FIGURE 10-2

The transmedia model showing the clean potato campaign plans

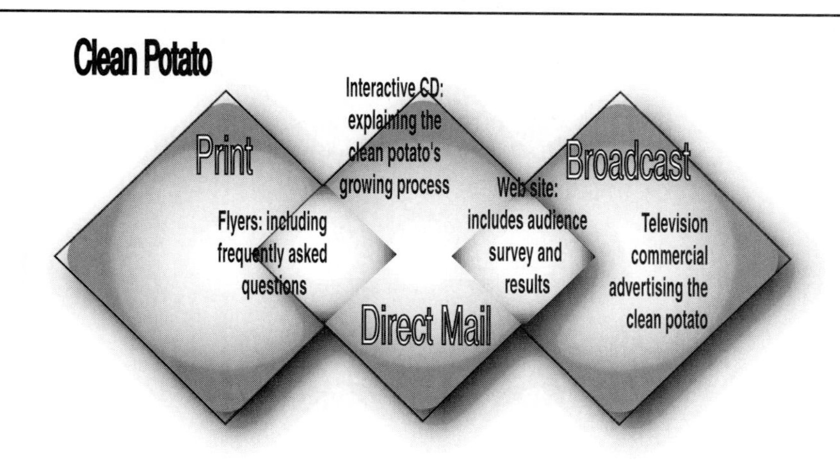

into account the general perceptions of their target audience. As we discussed in Chapter 4, a media arts campaign involves creating a positive commercial presence for a client. Steven Heller, senior art director for the *New York Times* observes, Graphic design is about interdependence and interrelatedness (Wilde, 2000); Heller is referring to the marriage between technique and concept—form and content. The form a design takes is inextricably linked to the perception of its contents. It is especially important to consider the layout, images, and color scheme, because these elements usually determine how audiences react emotionally to the product or service being offered. The content of the copy and the nature of the service or product should be inherent in the choices of design structures for the transmedia solution. For example, studies have shown that, the more negative space in a layout, the more an audience will categorize the product or service as high-end or sophisticated. This is because abundant negative space provides a respite for the viewer—a rest area for the eyes. Intemperance with graphics promotes a feeling of tackiness, but open areas, whether they are in a design layout, an interior of a home, or the great outdoors, trigger emotional responses of richness and tranquillity, as well as a natural connection to the unending vastness of the universe. As abstract as this may seem, the audience's emotional responses to media are vital to the success or failure of the media communication. In our clean potato example, choosing a limited palette of earth tones with accents in gold text may convey a feeling of natural richness. At the same time, two-thirds of the layout is negative space, and the final third is a deceptively simple image of a single mashed potato dish with scrumptious dressings. The disclaimer about the biologically altered veggie could be buried in the colors of the layout background, allowing the audience to focus on the emotional triggers of hunger and beauty. See Fig. 10-3. Again, it is wise for producers and designers to research the demographics and psychographics (values, beliefs, attitudes, and behaviors) of their target audience before making design, layout, color, and audio/visual decisions.

FIGURE 10-3

A cluttered text layout versus an uncluttered text layout

A Technology Selection

Following the journey of our clean potato through the use of Web site surveys, newsgroup postings, and dedicated Internet chat rooms, researchers for the advertising company have identified a target audience demographic and psychographic. In addition, the researchers have compiled a list of desirable and undesirable terms and images related to the clean potato product, according to audience responses. Relying on this research, the design group with the ad company decides to use four main types of media to promote the clean potato: a TV commercial, a Web site, an interactive CD, and a printed direct mail flyer.

First, the company will develop a TV commercial promoting the nutritional and extra teeth-cleaning virtues of the potato. The ad agency may have an internal production department or go to an outside production company or broadcast station, depending on the deal offered by each. Broadcast stations are used a great deal, because the stations offer airtime tied in with producing the commercial. Either way, the process discussed in Chapter 6 will occur, and the idea will be taken from a story treatment to production using digital film technology. Once a script is finalized, a production crew will shoot and edit the commercial with the producer. The final version of the commercial will be placed in the desired analog, digital, or Web format for airing.

The digital version of the commercial can be embedded into the clean potato Web site. The target audience research indicates that most visitors to this type of Web site are presently using DSL (Digital Subscriber Line) or cable modem connections to the Internet, warranting the inclusion of high bandwidth (high-speed), rich media in the clean potato Web site. Consequently, the digital video must be resized, cropped, and color corrected according to the file CODEC (coding/decoding) format for optimum delivery. The video will be compressed to fit the parameters of a Web page. It is decided to include both a simple upload and a streaming version of the video. The simple video will have a link explaining to the user that, if chosen, the entire movie must be downloaded to the user's hard drive before it begins playing. Ideally, small-format videos work in this environment. The second option is a streaming video; using the proper embed tags in HTML, a user with a newer browser and the right QuickTime plugin will be able to view a movie after a certain amount of data streams, or downloads, onto his/her computer. While playing the initial footage, the browser will continue to download the rest of the movie. Reformatting the video to standard formats such as Apple QuickTime and RealVideo ensures that the majority of viewers will already have the third-party plugin bundled with their browsers to play the media or can easily get the free downloads from the manufacturer's Web site.

As a complement to the digital video, various GIF animations produced in the bitmapped graphics animation program Adobe ImageReady are used in the site to explain the agricultural growing process of the clean potato. Visitors to the site can have their questions answered by the manufacturer through both e-mail and FAQs (frequently asked questions). The Web site also hyperlinks to other successful biologically altered veggies' sites. These companion sites have been contacted by the ad company to make sure there is a complementary link back to the clean potato site from the auxiliary site. Third, the digital version of the commercial is to be included on a trailer CD, which will accompany the direct mail flyer. At this point in time, it is safe to assume that most users of the Web have a CD-ROM drive fast enough to play digital video CDs. The video is compressed to the acceptable standard for CDs (320 x 240 pixels per inch), which is one-quarter the size of full-screen video, arranged on the hard drive and burned using a CD-RW (CD read writable) drive. There are a number of CODECs built in to QuickTime that are ubiquitous for compressing video to CD-ROM. One type is Cinepak; unfortunately, it is based on a lossy compression method, but it is the best choice as a compression format for new and old machines. On the other hand, MPEG1 CODEC produces the highest-quality output, which tends to be the recommended compression format for faster computers with high bandwidth connections. The designers decide to go with Cinepak and burn the video to a 650 MB standard CD, which costs about $1.00 per disc. Focusing on scrumptious images of the clean potato in various recipes for the full-color flyer, the ad company creative professionals work to bundle the CD to grow out of a natural extension with the printed information. All parts of this media campaign are visualized at the very beginning of the design process. Obsolete notions about the lines between media have been discarded. The clean potato is ported to various media outlets using component print, audio, video, and animation elements, which are in essence interchangeable parts of the whole.

Putting It All Together

We have seen how the tools of technology have radically changed the way producers, designers, advertisers, marketers, and broadcasters create media communication today. Rapidly changing technologies inundating the media industries have brought about a generation of producers and designers who are mandated by necessity to continually update their technical repertories. Is this a feasible model for the future?

Philip Michaels, who gives a unique insight into the time-collapsing nature of communication technology invention, published a pertinent time line in an issue of *MacWorld* magazine. He relates the following (Michaels, 2001, p. 28). The first full-length Technicolor movie, *Becky Sharp*, was developed in 1935, while the first color TV broadcast of *Dragnet* occurred in 1953. Further, the first color Macintosh, Mac II, came onto the market around 1987, while the first *New York Times* front page in color occurred in 1997. Finally, the first color palm device became a reality in the year 2000. As apparent, the time span between the airing of the first color movie and the first color TV broadcast was eighteen years; however, this pattern of times starts to collapse as we move closer to the present. New ways to develop and deliver media are seemingly daily occurrences. Is the content driving the use of these technologies, or is it the other way around? Many media producers and designers believe that the technology is driving the content. If a photographer cannot get an actual photograph of two celebrities together in the same place for the morning run of a newspaper, then he/she can electronically image-edit two individual shots of those celebrities to look as if they were together. There is a danger in using technology to alter reality when reality is what one is reporting.

Further, media designers and creative advertising people have the ability to manipulate imagery so completely that it is virtually impossible to detect any alteration. This is an awesome power. One of the most tragically poignant examples of reality almost being mistaken for special effects occurred in the TV clips of the airplane attacks and subsequent collapse of the World Trade Center in New York City on September 11, 2001. Many TV viewers initially thought that the video of the tragedy was some kind of "special effect." They had seen so much of that kind of imagery in recent movies. The tragedy was horribly real, but most of the people who had witnessed the actual incident and had viewed the ruins in person realized that the devastation lost much of its impact on the TV screen. The TV images could not capture the reality. Is it because media producers we have become so adept at manipulation through technological tools that they have oversaturated their audiences? Will there be a backlash? Could the media communication technology pendulum swing back toward the use of nontech tools? Witness the reluctance of the *New York Times*, considered by many to be the premier communication periodical in the United States, to adopt color technology for ten years, even though national newspapers adopted the technology for color newspaper printing in 1987 (specifically, USA *Today*). It is inevitable that the technology that has made producing TV, movies, audio/video, print publications, and Web publications easier will continue to be upgraded. However, regardless of the state of the technology, the ethics and responsibilities of the designers and producers of media are steadfast. What remains to be seen is whether or not as media consumers we demand accountability from them.

MEDIA DESIGN COMMUNICATION

Gordon Moore's famous law of technology, in which he predicted that progress in electronics—namely, computer speed and capability—would double every eighteen months is conservative concerning the current rate of increases in microchip technology. Bandwidth is being installed twice as fast as Moore's law. Rick Stevens, a professor of computer science at the University of Chicago and Argonne National Laboratory, predicts that, by 2040, processor chips might be the size of atoms at the present rate of upgrading. Media communication is generally predictable in that technology trends, which solidify into standards of industry and practice, can pinpoint future "killer applications." An example of this is the revolution of the printing industry by desktop publishing. Before the mid 1980s, digital typesetters were available, but their interfaces were complicated and their cost was prohibitive. In other words, these digital typesetters set type faster than traditional copyfitters and paste-up artists, but the technology could not be translated to the masses. It was once again proprietary—as difficult to learn and cumbersome to produce as the stone lithographic methods of the early twentieth century. Two technological milestones changed these parameters: Adobe systems created a layout program, called PageMaker, with an elegantly simple graphical user interface (GUI) and a powerful font generator, while personal computer printer manufacturers perfected a low-cost ink-jet printer capable of typeface recognition and crisp color output. These developments combined to alter the way print communication moguls conducted business because of the massive adoption of these technologies by nonindustry sources. Now, everyone can print a professional-looking full-color newsletter or flyer without leaving the office. The printing industry was compelled by these technologies to change or face obsolescence. Adobe PageMaker was the "killer app" that began the revolution that continues today. Inventors and manufacturers of new media technologies in software and hardware continue to search for the newest "killer app." Obviously, it is not possible to be absolutely sure which technologies will survive and become industry standards, like Adobe PageMaker; only time will tell.

COMMUNICATION EDGE TECHNOLOGIES

As difficult as predicting the next "killer app" may be, there are some emerging hardware and software technologies that are valuable for their present innovation and their future usefulness. The following is a discussion of some interesting "edge technologies," as well as some predictions about those that are presently having an impact on the way we communicate.

In the publishing world, e-books, books that are published in electronic form only, are inherently different from their paper counterparts. E-books are transitory, while paperbound books are permanent. Paperbound books represent a long line of editorial revision from the initial author, through reviewers and editors, and finally to the reading public by way of future editions. E-books can be published by one person and viewed by potentially millions without so much as one other human editing its contents. These digital books can be revised on the fly, added to, deleted from, or simply republished as part of a different compilation. The speed, cost-effectiveness, and ease of getting an e-book published are significant advantages over a paperbound book. Eventually, e-journals, e-books, and e-zines will supplant traditional hardcopy publishing.

Hardcopy book libraries are fast becoming anachronisms. Online libraries offer, for a nominal fee, full electronic texts of the classics. In addition, these libraries are adding thousands of books to their collections at exponentially increasing rates. A user can access the books from the online library, print out the text or read it on his/her desktop system, or download the book to a hand-held microprocessor specifically designed for reading electronic texts. These e-book devices are easy on the eyes and can hold hundreds of books, which can be purged after the user is finished reading the text and reloaded with a new set of tomes. In addition, rather than static text, as in a paperbound book, these books have dynamic interfaces. The readers can simply click on a term in the copy to see an explanation about the word. These e-book devices can be linked to a wireless Internet connection, permitting the user to view applicable Web sites linked to the text.

In mass communication outlets, including magazines and newspapers, traditional publishing practices are being challenged. Target audience markets are being dispersed into concentrations of various niche markets. Rather than blanketing a general target audience, such as women between the ages of eighteen and thirty-four, and producing an expensive glossy magazine, publishers are narrowing their focus to pinpoint more specific targets, such as Asian women eighteen to twenty-four years old with an income between $20,000/year and $45,000/year, living in Chicago, Illinois. Publishers can economically print full-color magazines for this type of niche market, primarily because of the revolution and democratizing of the graphic arts (printing press) industry. In the near future, a print publication that does not cater to a minutely specific audience and make its information available in print and a complementary online publication will be in great danger of obsolescence.

As bandwidth increases, the line between publisher and user will blur, because faster connections with veritably instantaneous feedback promote interactive publication, which is continually being manipulated by electronic updates and user-driven design. For example, corporations are finding creative ways to glean information from their present and future customers by offering gamelike interfaces for their Web portals and Web sites. As customers answer marketing questions and give personal information, the companies offer invitations to electronic gaming arenas or online contests. This dynamic activity is compiled and republished almost as fast as it is being processed. The users are equal partners in determining the structure and design of the interactive publication, which is constantly changing. The company is benefited by increased customer satisfaction as well as marketing information, and the customer is rewarded by seemingly free entertainment.

In the arena of graphic communication, digital camcorders with 3-D imaging software can create animations of rotating 3-D models and demonstrations describing any subject matter available in traditional media. The interesting aspect of this hardware technology is its ease of use. Historically, attempting to describe a 3-D object in 2-D space required high-end mathematics and programming expertise. This hardware/software combination allows a user to simply use a scanner (the camcorder) to input the visual information with minimal effort from the human producer. Multipurpose digital cameras that take high-resolution still shots as well as high-resolution video can offer three modes: video, photo, and progressive photo (captures photos to a multimedia card). Hybrid sheet-fed color scanners that incorporate paper management software enable the user to manage all electronic documents and images by creating drag and drop thumbnails of more than thirty file types. This drag and drop capability allows users to grab a file's icon, drop it onto the link icon, and send it to a variety of applications or hardware devices. These scanning devices and others are making the input of complex visual information completely painless and ultimately user-friendly. For example, electronic whiteboards with LCD projectors allow teachers, trainers, and the like to import and browse through a Web site while scribbling information on the monitor. These notes can be captured in screen shots and posted on additional Web sites, compiled into an e-book, or projected as a digital lecture to present in front of a live audience. In the end, these media data can be ported to multiple output media, infusing this process with infinite flexibility, especially for non-technical professionals.

Image creation and editing tools are becoming even more integrated, incorporating components of vector, bitmap, animation, and Web tools. Jasc Software's Paint Shop Pro 7™ is presently very popular because it is affordable while including an image editor, vector drawing tools, special effects, image optimization, a built-in image browser, and GIF animation all in one package. In the Web animation region, Flash technology is replacing online banner advertisements. According to a recent DoubleClick study conducted by Diameter, Flash increased branding metrics by 71 percent for three different-sized ads. Macromedia Flash Player software is distributed to more than 97 percent of online users. In the same vein, freeware and shareware are booming because, like the free Flash Player, a user can tap into the newest or most popular software and download it for use on his/her own system. However, most online professionals predict that the era of a free World Wide Web is coming to an end. Already, thousands of Web sites are requiring fees for access to more than preliminary information. In addition, most Web surfers are ignoring anything on a Web page that looks like an advertisement, giving rise to the practice of paid search engine placement by companies such as American Express, eBay, and Office Depot. Even in their infancy, online media environments have grown from a text-based research tool to an "edutainment" communication experience.

Expensive animations, such as those produced by the Disney Studios are making way for much smaller, lower-cost electronic productions. The computer has pushed traditional cell animation techniques into electronic automation workflow, with a noticeably smaller animation artist pool. The most popular professional 2-D animation suites in the Western world, Cambridge Animation Systems' Animo™ and Avid's Softimage Toonz™, provide expert and novice alike with the ability to build complex animations in a fraction of the time and cost that it used to take in a traditional studio. It is now possible for one or two people to open a lucrative animation studio and create professional-level animations. Similarly, in the 3-D world, PC-based animation suites, such as NewTek LightWave 3D™ propel previously low-end, low-cost systems into the competitive, professional broadcast markets. However, media communication produced by these new techniques and tools is processor-intensive and requires enormous amounts of disk storage capacity.

Finding innovative ways to alleviate these drawbacks reminds one of an earlier day in computing history. As software and hardware connectivity become more standardized, the individual boxes with programs loaded to isolated hard drives could make way for a resurgence of a neo-mainframe system. Software programs, regardless of their complexity, size, and processing requirements, will be available by accessing the main hard drive from terminal-like network boxes. The difference between yesterday's and today's mainframe is the solid foundation in backup technology. In other words, if the main system "goes down," an alternate system immediately kicks into gear and fixes the problem on the main system. See Fig. 10-4. Many communication technology professionals are looking to this type of network technology, because the cost of maintenance, human support, and error handling in the present PC-oriented system is prohibitively expensive and difficult to manage.

CONCLUSION

Authoring tools will eventually render the programming components for creating multimedia totally invisible, allowing producers to use multimedia objects as customizable drag and drop applications. Macromedia Director already provides producers with a constantly evolving tool set. Along with easier to use and lower learning curve multimedia programs, the distinction between virtual reality and actual reality is being explored. The "gamers," or edge-loving technology users, were the first adopters of emerging communication in virtual reality. Worldwide virtual reality games—EverQuest™, Ultima Online™, and Asherons's Call™—in which users concoct entirely new virtual identities and play with participants located all over the world, are seeping into our collective experiences. Underlying this seemingly trivial pursuit is the idea that communication technology has granted users the ability to build networks of communication untouched by actual human interaction. In animations such as *Final Fantasy*, the computer-animated characters are so real that they essentially confuse natural reality completely. The computer-generated actors actually have human talent agents, attesting to the fact that human communication is being altered.

The possibility of electronic communication usurping our present mass communication venues is real; however, a technological readiness quotient may be the key to the actual global implementation of virtual communication and e-business models. The Economist Intelligence Unit e-business forum, an Econ organization, recently published a country-by-country ranking of e-business readiness of sixty countries. The survey established four basic categories analyzing infrastructure, social and cultural

FIGURE 10-4

A comparison between a personal computer environment and a supercomputer environment

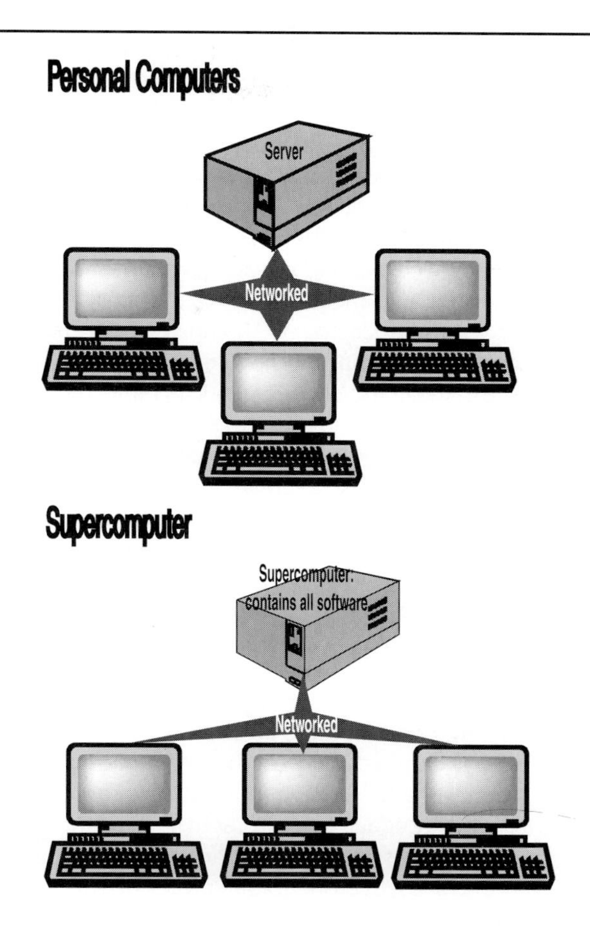

factors, legal environment, and supporting e-services. The top five countries, in order of most e-business ready, are the United States, Australia, the United Kingdom, Canada, and Norway, with the five least ready being Nigeria, Kazakhstan, Vietnam, Azerbaijan, and Pakistan. Not surprisingly, close to 60 percent of the countries on the list fall into the realm of those at risk of being left behind because they face major obstacles, primarily in the area of connectivity. This e-business readiness scenario is a valid example of the roadblocks that need to be addressed before virtual communication becomes commonplace.

As one can see, the media technology tools that enable the creation of e-communication content are only one piece of the multimedia communication pie. The remaining wedges comprise journalists, graphic communicators, media designers, e-producers and generally all creative professionals skilled in the delicate balance among technical tools, transmedia venues, and communication content. The democratizing of broadcast and published communication through the digitizing of media production tools has created exciting production environments as well as potentially frightening human communication consequences. Is communication possible through media technology alone, without human interaction? Presently, computers can be programmed to learn from processing mistakes and to adapt to a multitude of environments. Is the mere fact that these processing units are not "walking around" as humans do the only remaining differences between humankind and machine? Our technology expertise through transmedia projects is rapidly improving, so much so that the lines between natural reality and virtual reality are blurring. Virtual reality (VR), totally abstract algorithmic-based design, can be hypnotizing in its realness. Communication professionals are tapping into VR in a myriad of ways: real-time tours of any imaginable vista that can be accessed online, virtual characters with apparently "human" personalities that can be built and manipulated by their creators, and the augmentation of reality—virtual and physical reality seamlessly combined. Augmented reality (AR) can be used as a mediator of communication across a network in a shared virtual reality space. An ancestor of AR can be identified in the early integration of animation with footage of human actors in films dating as far back as the 1950s—remember when Jerry, the cartoon mouse of *Tom & Jerry* fame, danced effortlessly with Gene Kelly? Augmented reality may provide new directions in the communication process in the future. Today, it is being used in movies, videos, animations, and graphics primarily to save time and money as well as to create visual effects that were unimaginable just a few years ago.

Our discussion about media design communication technology throughout this book has spanned media harmony, messages, perception, visual technology, audio technology, and the online world. It is appropriate that we summarize these practices by using an acronym involving a basic nonverbal communication—a smile. The crux of media design can be defined in five distinct areas: *s*peed, *m*erging, *i*nterface/interactivity, *l*anguage, and *e*conomics. First, media design has changed dramatically with the ability to create, send, and distribute messages instantaneously—speed. Second, the convergence of media hardware has produced a whole new venue of communication possibilities—merging. Third, the simplifying and streamlining of the media tools have made it possible for the public to be both producer and audience—interface and interactivity. Fourth, the democratizing of proprietary trade jargon allows laypeople to enter the professional media development arena—language. Fifth, the low cost of previously professional-level software packages and hardware suites has made media design development accessible to the masses—economics. Although new communication technology is available to practically all who have access to a personal computer, the essence of successful media design lies in the hands of the producers, not in the hardware or software. Beyond technical prowess, the communicated message is effective only if it is understood.

BIBLIOGRAPHY AND SUGGESTED READING

Craig, James. (1983). *Graphic Design Career Guide*. New York: Watson-Guptill.

Edyburn, Dave L. (1999). *The Electronic Scholar: Enhancing Research Productivity with Technology*. Upper Saddle River, NJ: Prentice Hall.

Farace, Joe. (1996). *The Photographer's Digital Studio*. Berkeley, CA: Peachpit Press.

Fishel, Catherine. (2001). Creativity. *Step by Step Graphics*, 17(4), 42–47.

Fisher, Robert V. (2000). Overcoming Obstacles to Multimedia for the Web. *Syllabus*, 14(3), 58–59.

Griffin, Hedley. (2001). *The Animator's Guide to 2D Computer Animation*. Woburn, MA: Reed Educational and Professional Publishing.

Ibanez, Ardith, and Natalie Zee. (1998). *HTML Artistry: More Than Code*. Indianapolis, IN: Hayden Books.

Joinson, Simon. (Ed.). (2000). *The Digital Photography Handbook*. London: Duncan Petersen.

Kerlow, Isaac Victor. (2000). *The Art of 3-D Computer Animation and Imaging*. New York: John Wiley and Sons.

Lewell, John. (1985). *Computer Graphics: A Survey of Current Techniques and Applications*. New York: Van Nostrand Reinhold.

Long, Ben, and Sonja Schenk. (2000). *The Digital Filmmaking Handbook*. Rockland, MS: Charles River Media.

Long, Phillip D. (2001). Trends: Digital Convergence and the Interchangeability of Bits. *Syllabus*, 15(1), 8.

Michaels, Philip. (2001, April). Black and White Blues. *Macworld*, 28.

Olsen, Gary. (1993). *Getting Started in Computer Graphics*. Cincinnati, OH: North Light Books.

Parker, Roger C. (2000). Built to Order. *Publish: Print, Internet and Cross-Media*, 15(7), 38–42.

Simon, Eric J. (2001). "Are e-Books Ready for the Classroom?" *Syllabus*, 15(2), 28–29.

Truckenbrod, Joan. (1988). *Creative Computer Imaging*. Englewood Cliffs, NJ: Prentice Hall.

Widman, Jake. (1994). *Dynamic Computer Design*. Cincinnati, OH: North Light Books.

Wilde, Richard and Judith. (2000). *Visual Literacy: A Conceptual Approach to Graphic Problem Solving*. New York: Watson-Guptill Publications.

Glossary of Technical Terms

Acrobat (PDF) Reader: A software program that exclusively displays portable document reader files.

Active mixer: A volume control unit that controls several input sources at the same time.

Adobe Go Live: A Web editor software program used to create Web pages and Web sites.

Adobe Illustrator: A vector-based software program used to create digital drawings.

Adobe ImageReady: A bitmapped-based software program used to create graphics and GIF animations for the WWW.

Adobe LiveMotion: A software program used to create animations.

Adobe PageMaker: A layout-based software program used to create desktop publishing for traditional print media.

Adobe PhotoShop: A layout-based software program used to create desktop publishing for traditional print media.

Adobe Premiere: A software program used to create and produce digital video programs.

AI (artificial intelligence): The ability of a computer to learn and think as human beings.

AIFF (Audio Interchange File Format): An audio file format.

Aliasing: The instance of jagged, or pixilated, edges that occur in enlarged bitmapped images.

Analog: A system that sends a signal of electrical impulses through the electromagnetic spectrum to television sets.

Animated elements: Elements that can be as simple as a button graphic changing color when it is clicked or as complicated as interactive splash page animation.

Animation: A series of pictures. Each frame is changed slightly from the preceding frame to produce the illusion of movement.

Antialiasing: A software convention that optically smoothes the jagged edges of a bitmapped image.

ARPA (Advanced Research Project Agency): A project that aided in the development of the Internet.

Asheron's Call: A world wide virtual reality game.

Aspect ratio: The width and length of a television screen.

Asymmetrical: The frame is balanced, but the elements within the frame are not similar.

ATM (Adobe Type Manager): A software program that allows non-PostScript printers as well as monitors to display clean, crisp text outputs.

Audio levels: How the volume quality of audio is measured.

Audio Vision: A Mac-based digital audio editing software program.

Augmented reality: Reality that is digitally created and does not exist in the real world.

Authoring: The production of a multimedia program that uses interaction, such as a tutorial for a computer program.

Authorware: A software program that allows the producer to create an interactive multimedia program.

AVI: A video file format for the Windows platform.

AVID: A computer used to digitally edit and create video and television programs.

AVID Softimage Toonz: A software program with the ability to build complex animations in a fraction of the time and cost it would take in a traditional studio.

Background: The area located in the upper part of a screen and the area located behind graphics or text on a screen.

Bandwidth: The space that holds the data or information carried within a certain frequency or over a transmitting line.

Beta: A 1980s videotape format, which was unsuccessful in the consumer market.

Binary: A numbering system based on 0s and 1s.

Bitmap: A rectangular array of picture elements (pixels) in a map, each of which is encoded as a single binary digit. Every pixel is mapped at a certain location (address) in small, square dots, which merge optically when viewed at a distance from the screen.

Blueline board: A layout board on which elements that print in black are physically pasted up on a main board for color separation purposes.

Broken link: A link on a Web page that does not lead to anything.

Browsers: Software programs that interpret HTML code and display them in a user-friendly format for access to the many Web sites on the Internet.

Burn: To record on a CD or DVD.

Calibration: The integration of different machines, so that the signal transmitting through them stays at a quality level.

Caller ID: A telephone's ability to identify the caller on a screen.

Calligraphy: A type of print.

Cambridge animation systems: A combination of a digital animation software suite of programs and a hardware turnkey system developed to make traditional cell animation more economical and easier to produce.

CD (compact disc): A disc that holds digital audio or video.

CD player: A machine that plays a CD.

CD-RW (compact disc read writable): A CD that can be played and recorded on.

CERN: The European Laboratory for Particle Physics, where in 1991 the World Wide Web was developed.

CG (character generator): A computer that creates electronic titles that appear on television screens.

CIELAB (Lab color): A three-channel international color mode, which was developed for the purpose of color consistency between digital devices. The channels represent lightness (L), the colors green to red (a), and the colors blue to yellow (b).

Cinepak: A compression format used for new and older computers.

Clip art: Digital pictures and graphics that can be used with many different computer programs.

CMS (color matching system): A way to match colors when using different technology.

CMY (cyan, magenta, yellow): The subtractive color primaries used in the printing industry to create all colors printable.

CMYK (cyan, magenta, yellow, and black): Black is normally added to formulate richer colors in printed pieces.

CODEC'S: The players or plugins used in the compression process.

Color to plate (CtP): A streamlined process whereby the film step in a color separation is eliminated and the digital files are translated directly to an offset lithographic printing plate.

Combined media: Multiple media that are integrated into one medium.

Commercial media: Media produced for the express purpose of generating revenue.

Commercial printing: Printing produced for the express purpose of generating revenue.

Communication technology: Media as vehicles/technologies that transmit information to other human beings.

Compression: Reducing the amount of data in a computer file by leaving out some of the information and basically squishing the file down in size.

Computer conferencing: Using computers to see and speak to several people at once.

Content: Information or a message being communicated.

Copyright: The legal right of ownership over original audio/visual material.

CPU (central processing unit): The part of the computer that controls its operations and computations.

Critical media analysis: Decoding media messages through a qualitative methodology.

Critical viewing: Becoming an active member of an audience by moving one's perspective away from the perspective created by the media.

Cross-media publishing: Using various media to push the same message.

CSS (cascading style sheets): A standard method of creating formatting instructions and saving them for use with all of a site's documents, much like style sheets in a layout program.

Cyber: An environment that does not exist in a real-world context but is created by communicating through the use of computers.

DAT (digital audio tape): High-quality audio tape that records information in a binary code.

Data transfer rate: The speed at which information flows through transmitting equipment.

DAW (digital audio workstation): A computer with audio editing software, such as Sound Forge or Pro Tools, which mixes together multiple tracks of audio into audio files to be used with video or alone.

DCS (desktop color separation): A file format for saving a CMYK image for color separation with the options for including spot color channels, alpha channels, and a low-resolution file for previewing.

Dead link: A Web button, text, or image that is supposed to lead to another Web page but does not.

Desktop publishing (DTP): The traditional print processing steps were simplified and significantly compressed by the ability of a software/hardware combination of tools, taking ideas from design to camera-ready in a series of simple steps.

DHTML (Dynamic Hypertext Markup Language): Expanded classic HTML at the 4.0 level, including layers, absolute positioning, and style sheets.

Digidesign ProTools: A software program that allows the recording, editing, and playing of audio.

Digital: The recording or transmitting of data in a binary code.

Digital BetaCam tape: A traditional videotape format that uses digital technology.

Digital editing: The ability to move media around in a nonlinear fashion to create a media clip or program.

Digital online workflow: The process of creating printed media exclusively through digital communication.

Digital video: Visual information that is recorded in a binary code onto a tape, disk, or hard drive.

Digitizing: Recording or transmitting media in a binary code.

Director: A person who controls a production crew and the overall look of a media program.

Dissolve: An effect in which one image fades onto the screen while another fades out.

Downloads: Transfers of data from location such as the Internet to a personal computer.

DPI (dots per inch): The number of pixels that make up an image.

Drag and drop publishing: Computer programs that give the user the ability to pick up various media and move them around a computer screen with the click of a mouse.

DSL (digital signal line): A dedicated digital telephone line that carries large amounts of data at a fast speed.

DV tape: A digital tape format.

DVCAM tape: A digital tape format.

DVCPRO: A digital tape format.

DVD (digital videodisk): A digital disk format.

E-books: Electronic books that can be read from a computer screen or specially designed compact screen.

Edit window: The screen window on which editing takes place.

Editor: The person who actually moves the media around in the edit window to create a media clip or program.

Edutainment: The combining of education and entertainment to produce a new type of program.

Electromagnetic spectrum: Energy waves on which audio and visual media are attached to transmit this information around the earth.

E-mail (electronic mail): Digital messages sent and received over computer networks.

Embed tag: In HTML code, the tag that embeds a plugin or an ActiveX control into a Web page.

Enhance TV: TV programs that use interactive Web sites as part of the show.

ENIAC (electronical numerical integrator analyzer and computer): The world's first electronic digital computer.

EPS (Encapsulated PostScript): A vector file format that contains PostScript code for the printer and sometimes an optional PICT or TIFF image for screen display.

Event handlers: Commands that trigger actions whenever a certain event occurs, such as the action of a mouse click triggering an animation to play.

EverQuest: A game in which users concoct virtual identities and play with participants all over world.

Fade: To dissolve one image out of or onto a screen.

Fiber-optic technology: Technology in which glass rods carry large amounts of information at a fast pace in the form of light.

Field-dependent: The characteristic of someone who needs multiple senses engaged to process information successfully.

Field-independent: The characteristic of someone who needs only one sense engaged to process information successfully.

Filmware cards: Removable circuit boards, which can be plugged into a CPU to enhance aspects of a computer system, such as video, audio, and speed.

Final Cut Pro: A digital video editing software program used by amateurs and professionals.

Flash: A Web-based software program for creating and animating vector art.

Flash-file: An animation file that was created by the software program developed by Macromedia.

Flexography: A type of printing process using a raised-image, flexible plate, which makes it possible to print on irregular surfaces, such as coffee mugs and aluminum cans.

Flow chart: A diagram displaying the various directions information may move in a production project.

Formal balance: Both sides of the frame are symmetrical and create an ordered, calm environment.

Frame: A computer or television screen or a frame of reference for a graphic design.

Gant chart: An organizational flow chart.

GIF (Graphics Interchange Format): A bitmapped format used mainly for hard-edged graphics on the Web, such as clip art.

Grand Alliance: A group of companies that got together to develop and win the digital system standard.

Graphic communication: The practice of using Gestalt perceptual theories through visual technology to produce effective designed experiences.

Graphic design: The attempt to persuade, to inform, or to inspire through visual communication.

Gravure: A type of printing process, which uses an etched copper cylinder or wraparound plate; the surface of the cylinder represents the nonprinting areas.

Gray scale: The various tones that can be produced using black, white, and intermediate percentages between the two.

GUI (graphic user interface): The design of a Web site or page and how the user moves through the site.

Hardware: The devices that input, output, process, and display data.

HDTV (high-definition television): A hybrid system of analog and digital television.

Hexadecimal equivalents: Print colors or decimal values that are translated into hexadecimal values (00, 33, 66, 99, CC, and FF), the color system used by Web browsers.

High-speed access: Dedicated transmission lines that carry data quickly.

Hinting: In a Flash animation, hinting can be used by tagging points on the artwork in the first frame of a shape to correspond with the same points in the last frame of the shape. This allows the animator to control the "look" of the frames between in a shape.

Homepage: The main Web page that holds the user's welcome and usually opens the Web site to the user.

HTML (Hypertext Markup Language): A set of layout codes added to ASCII text that specifies the way the text is displayed.

HTML Composers: Software programs used to edit Web sites and pages.

Hue: A color.

Hybrid medium: One medium created from two or more media.

HyperCard: An older software program used to create interactive tutorials.

Hyperlinks: Buttons, text, or images that send the user to other Web pages or sites.

Hypermedia: The linking of text to other text (hypertext) but also the network of media elements connected by links on the WWW.

Hypertext transport protocol (HTTP): The coding protocol that allows computers to talk to each other by transferring HTML files on the Web.

ICC (International Color Consortium): A group of experts that describe standard color behaviors for each device used in the digital printing process.

Icon: A visual symbol used in computer interfaces.

Illustration: A visual, such as clip art, a drawing, a chart, or a diagram.

Imovie: Digital editing software used by the consumer market.

Index page: A Web page that provides information about the Web site's subject as a whole and becomes the portal for all relational subjects in the site.

Informal balance: Opposing compositional elements are asymmetrically arranged and yet are balanced, creating a more dynamic environment.

Information overdose: A person receives too much information and cannot process it all.

Information processing: Taking in data or information and organizing it to be used for various purposes.

Input device: Equipment, such as a digital camera, that puts information into a computer.

Interactivity: The ability of people to communicate back and forth with each other using a computer.

Interface: The overall screen design of software programs and Web pages.

Interlaced: A television scanning process in which odd, and even lines on a TV screen are scanned.

Internal microphone: The microphone that is built into a computer.

Internet: A vast, worldwide group of networked computers.

Internet service provider (ISP): A company or an organization that supplies users access to the Internet through the company servers.

Interpolate: To make a logical guess, as a computer processor does when adding the "extra" pixels needed to make an image larger.

Isomorphic correspondence: Audiences project their emotional reactions into a visual communication.

IT8: The most common target used today in color profiles for various input and output devices.

JavaScript: A scripting language, which specializes in controlling Web browser processes, most often through event handlers.

Jazz disk: A disk that can hold up to 1 gigabyte of data.

Job definition format: A description of a print production in the process of online printing.

JPEG (Joint Photographic Expert's Group): A bitmapped file format that also incorporates compression.

Junk communication: Messages that use every bell and whistle that technology has to offer but that pay little attention to content.

Keyline: In paste-up technique layouts, the alignment of color separations, tool lines, artwork, and text.

Killer app: A software application that makes all other applications in that subject area obsolete.

Kinesthetic projection: Extending an audience's responses to a visual beyond the physical limitations of the graphic.

Layout: The placement of visual text and images with the intention of communicating a specific message perceived by an audience within given limitations.

Lighting: Using lights to photograph an image to create the illusion of three dimensions and to set a mood, as well as to provide a strong compositional element.

Linear editing: Placing media clips in a sequential pattern by using analog equipment.

Links: Buttons, images, or text that can send a user to another Web page by clicking on it.

Lithography: A type of printing technique using a flat surface. The printing plate is treated chemically and holds ink because the printed area is receptive to oil-based ink but not to water.

Lossy compression: The loss of bits of information each time files are closed or opened.

Mac (Macintosh): The computer that runs on the Apple operating system.

Macromedia Director: An authoring software program used to create interactive multimedia.

Mailing lists: E-mail addresses saved under one address, so that the user can send out a message to several people at once.

Mailto: Forms that provide instant user feedback by means of e-mail links.

Master rights: Ownership rights of companies that make physical recordings of media.

Mechanical rights: The manufacturer's rights of distribution.

Media: Vehicles and technologies that transmit information to other human beings.

Media arts: Graphic communication disciplines converging to create a new hybrid area of study at colleges and universities.

Media blitz: Advertising placed in several media to get the largest audience.

Media Cleaner 5: A software program that converts media and files to streaming files.

Media convergence: The ability to use one medium that has the function of many media, such as playing a video on the Internet.

Media harmony: The use of different types of media to design effective communication.

Media-neutral: Messages that can be produced in practically any venue (Web, video, TV, print) without losing their impact.

Media 100: A hardware digital edit suite and a digital editing software.

Media-rich: Using many different media in a message.

Media stimuli: Media that capture audience attention.

Media text: Visual or audio material.

Metacommunication: Communication about communication.

Microchip: The semiconductor transitor that made digital technology feasible.

MIDI (musical instrumental digital interface): The audio file standard developed in the 1980s for electronic musi-cal instruments and computers to talk to each other.

MIME (multipurpose Internet mail extension): The part of an e-mail program that can read HTML formatting.

Modem: A telephone or cable device that physically hooks up a computer to the Internet.

Monitor: An output device that displays information.

Mosaic: The first icon-driven browser for the Internet.

Mouse: An input device.

Movable metal type: A kind of type used in early printing presses.

MPEG (Motion Photographic Expert's Group): A sibling to JPEG, except that it is the format for motion graphics and video.

MP3: An audio file that can be uploaded and downloaded easily.

MP3 player: A device that holds and plays back MP3 audio files.

MS Front Page: A Web-editing software program.

MS Internet Explorer: A Web browsing software program.

Multimedia: Many media used at once or a medium with interactivity.

Napster: The company that created the software program to download audio files easily from the Internet.

Narration: Audio heard over video clips, also known as voice over.

Navigation bar: The row of icon buttons used to move through a software program.

Netscape Communicator: A software program used to browse the Internet.

Netscape composer: A component of Netscape Communicator, used to create Web pages.

New technology: Devices or software programs developed recently that give the user more abilities for multiple activities.

Newsgroup: A group of users who share a common interest; they post messages to a shared server, which sends them to the individuals' e-mail accounts.

NewTek Lightwave 3D: A PC-based animation software program.

Nonlinear editing: Digital editing, which allows the editor to change the order of media clips spatially instead of from beginning to the end.

Non-PostScript: Unable to interpret PostScript language.

Now-loading: Information and data are moved in realtime from another source, such as the Internet to a personal computer.

NTSC (national television standard color): The U.S. standard television system for transmitting a signal to stations.

Offline editing: Editing a media program with basic equipment.

Offset lithography: A lithographic process that is the most common type of commercial printing process for traditional paper-based media.

Online editing: Editing a media program with high-end, high-quality equipment.

Opaque: Solid.

Optical resolution: The smaller of the two numbers used in describing a scanner's dpi—600 x 1,200 dpi indicates an optical resolution of 600 dpi.

Optimizing: Raising the overall level of audio as much as possible without distorting the sound before compression.

Output device: Equipment that produces information from a computer, such as a printer.

Painter: A software program that mimics the act of painting by creating digital paintings that emulate natural media, such as oil paint and watercolor.

Painting: Creating a picture using digital brush strokes.

PaintShop Pro 7: A very popular software program that is affordable while including an image editor, vector drawing tools, special effects, image optimization, a built-in image browser, and GIF animation all in one package.

Passive mixer: An audio control unit that can input and output microphone cables to increase the number of microphones being used.

Paste-up: An old-fashioned term used to describe the process of pasting photographic galleys of text and artwork elements onto a color separation board.

PC (personal computer): Hardware that uses a Windows-based system.

PDF (portable document file): A file format that can link and/or embed bitmapped images, fonts, and vector images into one file.

Performance rights: The composer's copyright.

Peripherals: Devices that attach to a computer, such as scanners and printers.

Pert chart: An organizational flow chart.

Photographs: Still images, captured on film or in digital format, that copy the look of reality.

PICT (Picture): A bitmapped file format.

Pixelization: The squaring of the pixels displayed or the ability to see the pixels instead of a crisp image.

Pixelparks: A slang term for a highly computerized and technical cluster of companies in a small geographic area, such as Silicon Valley in California.

Pixels: On a computer screen, the smallest visual element, which is essentially a square dot.

Plugin: A tiny helper application that enables dynamic media, such as QuickTime player.

PNG (Picture Network Graphic): A file format created to replace the CompuServe GIF and JPEG.

PostScript: The file type in which both the screen description and the outline print description are embedded.

PowerPoint: A presentation software program used to create digital presentations.

Preproduction: The steps taken before production to create a media program, such as creating a storyboard and a script.

Print collateral: Printing parts of books or one book for one customer.

Processing speed: The time is takes a computer to compile data.

Producer: A person who coordinates multiple activities simultaneously and is in control of an entire media project.

Production: The actual shooting and editing of media to create a program or a clip.

Production assistant: A person who helps with audio, gripping, and any extra activities needed to make a production flow smoothly.

Prototype: A clickable model, with placeholders for text and other media, that simulates the look and feel of a finished Web site.

Pulling: Seeking out and gathering information such as searching the Internet.

Qdesign's Music Pro 2: An audio compression software program.

QuickTime: An audio file format.

QuickTime Player Pro: A compression software program.

RAM (random access memory): The short term memory of a computer.

Readability: The ability to see graphics clearly.

Real time: The event is happening in the present.

RealAudio: A program that plays audio files.

RealVideo: A program that plays video files.

Removable disks: A disk drive that can be added to or removed from a computer.

Reveals: Animation effects in which there are two images in the window and one progressively masks the other.

RGB (red, green, blue): A color mode representing the additive set of hues (white light) that makes up the electronic palette of movies, TV, and computer monitors.

Rollover events: Animation effects in which the icon changes when the mouse moves over the top of it.

Root folder: The highest folder in a hierarchy (directory) of folders for uploading to a server.

RTF (Rich Text Format): A vector format used to transfer text from one application to another.

RYB (red, yellow, blue): The pigments used by artists and designers to create the spectrum of colors for paintings, illustrations, and paper-based graphics.

Safe zone: The area on a computer or television screen to place graphics so they do not go off the edge of the frame.

Sans serif font: A category of typefaces that typically display an even stroke weight, no stress, and no serif, or "chiseled," edge.

Saturation: The intensity of a hue.

Scanner: A device that reads hard copies of images and text and converts the images and text to a digital format.

Screenprinting: A printing process forcing ink through a screen stencil onto a printable surface, also referred to as serigraphy.

Script: The written description of a production, including audio and video directions as well as shot and audio descriptions.

Script editing: The changing of programming language.

Scroll: To move down or across a Web page.

Semiconductor chip: The microchip that made digital technology feasible.

Serif font: A category of typefaces that typically display a thick-to-thin varying stroke weight, with a stress, and including a "chiseled" edge.

Servers: Computers that communicate with other computers to create a network.

Shareware: Software that is freely shared on the Internet.

Shockwave Flash format: An animation file format.

Site navigation: How a producer wants a user to move through a Web site.

Site storyboard: A series of frames that display the look of each Web page.

Site tree: A flow chart that starts with a trunk, the homepage or index page, and branches out to the rest of the pages in the site.

Sketchpad: An early software program that allowed users to "sketch" vectors on a computer screen to be printed to a plotter.

Software: Applications or programs that work on a computer, performing several different functions, including running the computer.

Sonic Foundry's Sound Forge: One of the dominant digital audio software programs.

Sonic Solutions: A digital audio software program.

Sorenson video: A low bit rate CODEC program.

Sound designer II (SDII): An audio file format.

Soundscape: A PC-based audio editing software program.

Special effects producer: A person who coordinates the creation and use of computer-generated images, audio, text, or transitions for a media production.

Spectrophotometer: A hand-held calibration device for measuring color consistency between computer monitors and printing presses.

Splash page: An animation or image that displays before a homepage in a Web site.

Stepper motor: The motor that moves the scanning bar in a scanner.

Stone lithographic: See *lithography*.

Storyboard: A visual representation of a script.

Streaming audio: Audio that plays while parts of the file are still being downloaded.

Streaming content creation programs: Software programs that download content while playing it at the same time.

Streaming video: Video that plays while parts of the file are still being downloaded.

Stylus: A device that allows designers to draw on computer screens.

Switcher: A device with several input sources that can cut, dissolve, or wipe among the sources.

Symmetrical: Based on identical elements reflected equally in a frame of reference.

Synchronization rights: Composer's rights if their work is used with another medium, such as videotapes or the Internet.

System 7 sound: An audio file format.

Tag: The coding protocol used in HTML.

Technology: An instrument or a tool that improves the function of human activities.

Telecommunication Act of 1996: Groundbreaking legislation that changed the way media are regulated in the United States.

Teleconferencing: Several people in different locations sharing the same phone line for a meeting.

Templates: Guides used to create word documents, Web pages, digital presentations, and so on.

Themography: A finishing process, which uses nondrying inks on offset presses. After the ink is placed onto the printing surface, it is dusted with a powder. It is then heated, and the areas of ink swell or raise in relief to produce an engraved effect.

Three-dimensional (3-D): A visual that has depth as well as width and length.

TIFF (Tagged Image File Format): A bitmapped file format that is fairly universal for creating lossless quality color graphics.

T1 line: A dedicated telephone line used to connect to the Internet.

ToolBook: An old authoring software program.

Transmedia: Publishing a message in multiple media outlets.

TRT (total running time): The length and timing of a media production.

True type: Computer fonts that do not include a PostScript printer description, thereby making them undesirable when developing a print project.

T3 line: A dedicated telephone line used to connect to the Internet; it carries more data than a T1 line.

Typography: The study of type shapes, history, development, and production; also the type used in a composition.

Ultina online: A worldwide virtual reality game.

Universal Resource Locator (URL): A programming language that is Web addresses.

Uploaded: Sent from a personal computer to a server.

Users: People who view, listen to, or interact with various media.

Value: The lightness or darkness of a hue.

Vector animation programs: Software programs that produce animations using mathematical formulas (vectors). Typically, they are scalable and significantly smaller in size than bitmapped animations.

Vector illustration programs: Software programs that produce illustrations using mathematical formulas (vectors). Typically, they are scalable and significantly smaller in size than paint-based bitmapped images.

Video camera: A device that converts images to digital or analog signals.

Videographer: A person who is in charge of creating the visuals of a production and for bringing the director's vision to life.

Video-on-demand: The feature offered by media companies that allows users to download digital movies to a television.

Virtual reality (VR): An environment totally created by computers and not part of the physical world, except for the equipment used to enter it.

Visual clarity: The ability of the viewer to move easily through a computer or television screen.

Visual diagram: A picture of the overall information flow on a Web site.

Visual literacy: The ability to decode visuals.

Voice mail: Verbal messages recorded and played back from telephones and computers.

Volume unit meter (VU): An audio meter that reads the quality of a signal being recorded or transmitted.

VRML (.wrl) Virtual Reality Modeling Language: Developed by Mark Pesce and Tony Parisi, a spatial description language that enables visitors to a Web virtual world to look into it and navigate through it from every possible angle.

WAV: An audio file format.

Wave Frame: A PC-based digital audio editing software program.

Web editor: A software program that allows the user to change programming language on a Web site or page.

Web portal: A collection of many applications. The advantage in a portal is that server-side (at the location of the Internet service provider) programs can run within a single common interface.

Web presence: Having a Web page or Web site.

Web publishing: Placing material, such as an e-zine or a Web page, on the Internet.

Web site: A series of Web pages built by using HTML.

Webcast: Broadcasting a media program on the Internet.

Windows Media: A software program that plays audio files for Windows-based systems.

Wipe: The overt changing of one image with another, using patterns, such as a diamond.

Wireless technology: Digital towers and satellites that transmit signals through the air without the aid of cables or telephone lines.

WWW (World Wide Web): A linking of computers through the use of HTTP programming language.

WWW Consortium (W3C): An international group of volunteers from Internet software companies to academics that oversees the upgrades and development of technologies for the WWW.

XML (Extensible Markup Language): The language that was created as a more flexible coding language related to HTML; it includes tags and functions that can be defined by the publisher and displayed through a set of style sheets linked to the document.

Zip disk: A disk that holds 100 megabytes of information.

INDEX

a

Acrobat (PDF) Reader, 62, 130
Action painting, 36
Active mixer, 68
Adobe Go Live, 59, 92, 122. *See also* Web design programs
Adobe Illustrator, 58, 123
Adobe ImageReady, 60, 123, 147
Adobe LiveMotion, 126
Adobe PageMaker, 55, 149
Adobe PhotoShop, 54, 56, 105, 123, 131
Adobe Premiere, 60, 87. *See also* Digital editing software
Africa, 132
AI (Artificial Intelligence), 10
AIFF (Audio Interchange File Format), 69, 71, 73. *See also* Audio file formats
Aliasing, 56, 137
Analog, 4, 7, 8, 47, 49, 51, 56–57, 68, 80, 85–86, 88, 144, 147
Analogous, 43, 102
Animated elements, 124, 148
Animation, 1, 2, 14, 34, 45, 47–49, 55, 59–62, 89, 91, 96, 99, 104, 109, 114, 117, 121–126, 129, 135–136, 144, 151–152, 156
Antialiasing, 56
Apple QuickTime, 69, 71–73, 78, 89–90, 93, 124, 140, 147–148
  See also Audio file format
ARPA (Advanced Research Project Agency), 8
Asheron's call, 152
Asia, 132
Aspect ratio, 82–83
Asymmetrical, 38, 39, 82, 100, 105–106, 120
ATM (Adobe type manager), 56
Audio file formats, 69–79
  AIFF, 69, 71, 73
  MP3, 9, 71–73, 78–79
  QuickTime, 69, 71–73, 78–79
  RealAudio, 71, 78
  SDII, 71
  System 7 Sound, 71
  WAV, 69, 71–72
  Windows Media, 71, 78
Audio levels, 71, 86, 110
Audio Vision, 70
Augmented reality, 154
Authoring, 14, 53, 60, 72, 74–75, 92–93, 109, 131, 152
Authorware, 109, 125
AVI, 90–91
AVID, 87
AVID Softimage Toonz, 152

b

Background, 36, 38–40, 63, 100–102, 120–121, 126, 134, 141
Bandwidth, 77–78, 108, 126, 134–135, 140, 147, 149–150
Beta, 143

Binary, 49, 56
Bitmap, 14, 55–62, 125, 136, 151
Blue, 43–45, 102, 132–133
Blueline board, 56
Broken link, 136. *See also* Dead link
Browsers, 14, 48, 114–115, 137, 147
Burn, 55, 60, 72–73, 91, 148

c

Calibration, 51, 138
Calligraphy, 102
Cambridge animation systems, 152
CCD, 85
CD (Compact Disk), 2, 9, 55, 60, 68–69, 72–73, 76, 78, 89, 91, 107, 139, 144, 147–148
CD player, 68
CD-RW (Compact Disk Read Writable), 148
CERN, 14
CG (Character Generated), 86, 89
CIELAB (Lab color), 138
Cinepak, 89, 148
Clip art, 2, 4, 14, 62, 103
CMS (color matching system), 51
CMY (Cyan, Magenta, Yellow), 43. *See also* Primary colors
CMYK (Cyan, Yellow, Magenta, and Black), 43, 50, 62, 138
Codecs, 72, 90, 94, 108, 140, 147–148
Color to plate (CtP), 51
Color trapping, 51
Combined media, 1, 2
Commercial media, 12
Commercial printing, 48, 50
Complementary, 43–44
Compression, 8, 62, 69, 72, 76, 78–79, 88, 108, 140
Computer conferencing, 8
Cool colors, 44
Copyright, 73, 75–78, 94–95, 106, 123–124
CPU (Central Processing Unit), 52, 70, 87
Critical media analysis, 25
Critical viewing, 23–29
Cross-media publishing, 50, 55
CSS (Cascading Style Sheets), 124
Cyber, 96

d

DAT (Digital Audio Tape), 68, 70
Data transfer rate, 73, 108
DAW (Digital Audio Workstation), 69–70, 79
DCS (Desktop Color Separation), 62. *See also* Graphic file format
Dead link, 136. *See also* Broken link
Desktop publishing (DTP), 14, 49, 51, 59–60, 101, 149
DHTML (Dynamic HyperText Mark Up Language), 122
Digidesign ProTools, 69–71
Digital editing, 60, 71
Digital online workflow, 51
Digital video, 8, 10, 55, 60–61, 80–81, 86–87, 95, 138–139, 144, 147
Digital video editing software, 60
  Adobe Premiere, 60, 87
  dpsVelocity, 87
  Final Cut Pro, 60, 87
  iMovie, 60, 87
  Media 100, 87–90
Digital videotape formats, 8, 86
  Digital BetaCam, 8, 86
  DV, 8, 90
  DVCAM, 8
  DVCPRO, 8
Digitizing, 88–89, 120, 153
Director, 80, 84, 95, 130
Dissolve, 91, 109, 136

Downloads, 5, 7, 9, 51, 71, 73, 78, 90, 92–94, 114, 121–122, 126, 139–141, 150–151
DPI (Dots Per Inch), 57, 114, 137
Drag & drop publishing, 151
DSL (Digital Signal Line), 147
DVD (Digital VideoDisk), 68–71, 87, 91

e

E-books, 150–151
Edit window, 70
Editor, 2, 14, 60, 75, 80, 84, 87, 95, 113, 122
Edutainment, 151
Electromagnetic Spectrum, 6–7
Email (Electronic Mail), 5, 113, 115, 124
Embed tag, 147
England, 133
ENIAC (Electronical Numerical Integrator Analyzer and Computer), 47
EPS (Encapsulated PostScript), 62. *See also* Graphic file format
Event handlers, 124
EverQuest, 152

f

Fade, 70, 75, 109
Fiber-Optic technology, 7
Field-dependent, 20
Field-independent, 20
Filmware cards, 52
Final Cut Pro, 60, 87. *See also* digital video editing software
Firewire, 88, 91
Flash animation, 60, 62, 124–125, 136
Flexography, 50
Flow chart, 50, 118
Focus group, 145
Formal balance, 82, 100
France, 11, 133

g

Germany, 11, 133
GIF (Graphics Interchange Format), 62, 91, 114, 117, 123–125, 131, 136, 147, 151. *See also* Graphic file format
Grand alliance, 7
Graphic communication, 11, 13, 32, 34, 36, 40–41, 44, 48–49, 52–53, 63, 112, 151
Graphic design, 12–15, 32, 35–36, 38, 47, 49, 50–52, 54, 110, 112, 122, 128, 135, 146
Graphic file formats, 61
  DCS, 62
  EPS, 62
  GIF, 62, 91, 114, 117, 123–125, 131, 136, 147, 151
  JPEG, 62, 114, 117, 123–125, 131
  MPEG, 62, 90
  PDF, 62, 130
  PICT, 62
  PNG, 62
  RTF, 62
  TIFF, 61, 131
Gravure, 50
Gray scale, 102, 110
Green, 43–45, 102, 132–133
GUI (Graphic User Interface), 149

h

HDTV (High Definition Television), 5, 7
Hexadecimal equivalents, 114

High-speed access, 8, 147
Hinting, 93–94
Home page, 54, 118, 128
HTML (HyperText Markup Language), 2, 14, 59–60, 72, 112–113, 117, 122, 134, 147
HTML Composers, 14
Hue, 41, 43, 60, 121, 132–133
Hybrid media, 34
HyperCard, 113
Hyperlinks, 59, 114, 117, 136
Hypermedia, 113–117, 124, 126

i

ICC (International Color Consortium), 51, 138
Icon, 70, 74, 92–93, 117, 135, 151
Illustration, 35, 43, 54–56, 99, 101–103, 114
Image editing, 14, 51, 53, 138, 148
Imagesetting, 51
iMovie, 60, 87
Index page, 118, 121
Informal balance, 82, 100–101
Information overdose, 134
Information processing, 17
Input device, 52–53
Interactivity, 48, 55, 116, 122, 126, 137, 150, 154
Interface, 13–14, 47, 52, 54–55, 60, 71, 73, 88, 92, 116, 122, 125, 149
Interlaced, 139
Internal microphone, 66
Internet Service Provider (ISP), 112, 117
Interpolate, 56, 137–138
Isomorphic Correspondence, 37
IT8, 138
Italy, 133

j

JavaScript, 124, 137
Jazz disk, 69
Job definition format (JDF), 51
JPEG (Joint Photographer's Expert Group), 62, 131. *See also* Graphic file format
Junk Communication, 1, 5, 128

k

Keyline, 51
Killer app, 149
Kinesthetic Projection, 37

l

Layout programs, 55, 131, 149
  Adobe InDesign, 55
  Adobe PageMaker, 55, 149
  QuarkXpress, 55, 131
Lighting, 85–86, 98, 108
Linear editing, 86–87
Links, 62, 93, 113, 115, 126, 136, 151
Lithography, 50
Lossy compression, 62, 148

m

MAC (Macintosh), 13, 62, 87, 109, 114, 130–131, 143, 148

Macromedia Director, 60, 109, 152
Mailing lists, 115
Mailto, 124
Master rights, 76
Mechanical rights, 76
Media 100, 87–90
Media arts, 48–49, 52–54, 63, 116, 146
Media blitz, 145
Media Cleaner, 72, 89, 93
Media convergence, 2, 4, 6, 9, 142, 154
Media harmony, 1, 3, 15, 154
Media-neutral, 51
Media-rich, 3, 8, 119, 126
Media stimuli, 34
Media text, 26
Microchip, 149
MIDI (Musical Instrumental Digital Interface), 71–72
MIME (Multipurpose Internet Mail Extension), 93
Mini disc (MD), 69, 87
Mix window, 70
Modem, 7–8, 78, 87, 147
Monitor, 68, 70, 87, 89, 98, 110, 114, 120, 126, 133, 138, 151
Mosaic, 14
Mouse, 1, 52, 70, 87, 98, 124, 138
MPEG (Motion Photographer's Expert Group), 62, 78–79, 90–91
See also Graphic file format
MP3, 9, 71–73, 78–79.
See also Audio file format
MP3 player, 68, 73, 77
MS Front Page, 59, 122
MS Internet explorer, 72–73
Multimedia authoring, 53

n

Napster, 9, 78
Narration, 65–66, 69, 91, 99, 107
Navigation bar, 121
Netscape communicator, 72–73
Netscape composer, 73
New technology, 8, 15, 77, 111, 142
Newsgroup, 115, 147
NewTek Lightwave 3D, 152
Nonlinear editing, 87–88, 95
Non-PostScript, 56
Now-loading, 126
NTSC (National Television Standard Color), 139

o

Offline, 50
Offset lithography, 50. *See also* Lithography
Online, 50–52, 59, 115–116, 131–132, 150–151, 154
Opaque, 57
Optical resolution, 137
Optimizing, 72, 151
Orange, 43–44, 102, 121, 133
Output device, 52, 57

p

Painter, 57–58, 62
Painting, 4, 11, 36–38, 43, 49, 53, 57–58

PaintShop Pro 7, 151
Panoramic shot, 86
Passive mixer, 68
Paste-up, 51
PC (Personal Computer), 13–14, 51, 70–71, 87, 92, 143, 149, 152–153
PDF (Portable Document File), 62, 130. *See also* Graphic file format
Performance rights, 76–77
Peripherals, 52, 144
PICT (Picture), 62. *See also* Graphic file format
Pixelization, 138
Pixels, 49–50, 56–58, 108, 126, 137–139, 148
Plugin, 60, 71, 92–93, 114, 121, 124–125, 136, 140, 147
PNG (Picture Network Graphic), 62. *See also* Graphic file format
PostScript, 51, 56, 60, 62, 130, 141
PowerPoint, 71, 90–91, 109, 131
Preproduction, 82–84, 95
Primary colors, 43, 133
  CMY, 43
  CMYK, 43, 50, 62, 138
  RGB, 43, 114, 138
  RYB, 43
Print Collateral, 14, 51, 54–55
Printing, 4, 11–13, 32, 34, 43, 48–51, 56, 113, 141
Processing speed, 8, 108
Production Assistant, 84–85
Prototype, 118
Pulling, 3
Purple, 44, 102, 132–133

q

Qdesign's music pro 2, 72
QuickTime Player Pro, 72

r

RAM (Random Access Memory), 8, 58, 87, 108
Readability, 101, 102, 110, 121
Real time, 78, 93
RealAudio, 71, 78. *See also* Audio file format
RealVideo, 90, 147
Red, 43–44, 102–103, 132
Registration, 51
Removable disks, 122
Reveals, 136
RGB (Red, Green, and Blue), 43, 114. *See also* Primary colors
Rollover events, 122
Root folder, 117–118
RTF (Rich Text Format), 62. *See also* Graphic file format
RYB (Red, Yellow, Blue), 43, 138. *See also* Primary colors

s

SADiE Soundscape, 70
Safe zone, 139
San serif font, 120, 139
Saturation, 43
Scandinavia, 133
Scanner, 51–53, 56–57, 104, 137–139, 151
Screenprinting, 50
Script, 82–84, 95
Script editing, 14
Scroll, 126
Secondary, 43
Semiconductor chip, 8
Serifs, 120, 139
Server, 73, 112, 114, 125
Shareware, 53
Shockwave Flash format, 60, 125

Site navigation, 116
Site storyboard, 118–119
Site tree, 118–120
Sketchpad, 13
Sonic Foundry's sound forge, 70
Sonic Solutions, 70
Sorenson video, 140
Sound designer II (SDII), 71. *See also* Audio file format
Special effects, 148–149
Spectrophotometer, 138
Splash page, 121–122, 124, 135
Split complementary, 43
Stepper motor, 137
Stone lithographic, 50, 149. *See also* Lithography
Storyboard, 82–84, 117–118, 136
Streaming audio, 59, 65, 78, 119
Streaming content creation programs, 14
Streaming video, 2, 59, 86, 93, 119, 140, 147
Stylus, 32, 52
Switcher, 86
Symmetrical, 37–39, 82, 100, 105
Synchronization rights, 76–77
System 7 sound, 71

t

Tag, 112–113, 122
Telecommunication, 6
Telecommunication Act of 1996, 6–7
Teleconferencing, 8
Templates, 14, 73, 102, 122
Tertiary, 43
Themography, 50
Three dimensional (3D), 109, 151–152
TIFF (Tagged Image File Format), 62, 131. *See also* Graphic file format
ToolBook, 109
Transmedia, 130, 139, 143, 145, 153–154
Transport window, 70
TRT (Total Running Time), 84
True type, 141
Typography, 11, 56, 129, 140

u

Ultina online, 152
United States of America, 45, 133, 149, 153
Universal Resource Locator (URL), 114
Uploaded, 94, 117, 147
Utility, 53, 56, 61
Utility programs, 61
  Cassidy and Greene's Conflict Catcher, 61
  Norton's AntiVirus, 61
  Retrospect's Backup, 61
  Semantics' Font Suitcase, 61

v

Value, 32, 42–44, 56, 102, 106, 120–121
Vector animation programs, 14
Vector illustration programs, 14
Video camera, 80, 85, 91
Videographer, 2, 60, 80, 84–85, 95
Video-on-demand, 7
Virtual Reality (VR), 9–10, 152, 154
Visual clarity, 100
Visual diagram, 118
Visual literacy, 32, 35
Volume Unit (VU), 68, 86
VRML (.wrl) Virtual Reality Modeling Language, 78–79

W

WAV, 69, 71–72. *See also* Audio file format
Wave Frame, 70
Web design programs, 59, 72–75, 92–93
  Adobe GoLive, 59, 72–75, 92–93, 122
  Macromedia DreamWeaver, 59, 60, 72, 92, 122
  Microsoft FrontPage, 59, 72, 122
Web editing, 53, 59, 73, 92, 117
Web editor,14, 59–60, 122
Web portal, 48, 115–116, 150
Web presence, 8, 15, 59, 115–117
Web Publishing, 14, 47, 50, 59–60, 112–117, 149
Webcast, 9
Windows Media, 71, 78. *See also* Audio file format
Wipe, 109, 136
Wireless technology, 9
WWW (World Wide Web), 14–15, 52, 59–60, 62, 112–113, 115, 119, 120, 122, 131–132, 134–139, 151
WWW Consortium (W3C), 59, 112

X

XML (Extensible Markup Language), 122

Y

Yellow, 39, 43–45, 102, 133